Shell Petroleum Development Company, the State and Underdevelopment of Nigeria's Niger Delta: A Study in Environmental Degradation

Shell Petroleum Development Company, the State and Underdevelopment of Nigeria's Niger Delta:

A Study in Environmental Degradation

Daniel A. Omoweh

Africa World Press, Inc.

P.O. Box 1892
Trenton, NJ 08607

P.O. Box 48
Asmara, ERITREA

Africa World Press, Inc.

P.O. Box 1892
Trenton, NJ 08607

P.O. Box 48
Asmara, ERITREA

First Printing 2005

Book design: Sam Saverance
Cover design: Dapo Ojo-Ade

Library of Congress Cataloging-in-Publication Data

Omoweh, Daniel A.
Shell Petroleum Development Company, the state and underdevelopment of Nigeria's Niger Delta: a study in environmental degradation / Daniel A. Omoweh.
p. cm.
Includes bibliographical references and index.
ISBN 0-86543-983-4 (cased : alk. paper) -- ISBN 0-86543-984-2 (pbk. : alk. paper)
1. Petroleum industry and trade--Environemental aspects--Nigeria--Niger River Delta. 2. Petroleum industry and trade--Government policy--Nigeria--Niger River Delta. 3. Shell Petroleum Development Company of Nigeria--History. 4. International business enterprises--Nigeria--Niger River Delta. 5. Niger River Delta (Nigeria)--Economic conditions. 6. Niger River Delta (Nigeria)--Environmental conditions. I. Title: State and underdevelopment of Nigeria's Niger Delta. II. Title.
HD9577.N53 N5463 2001
363.738'2'09669--dc21

2001005449

To my mother, Echigeme Omoweh, 1910–2000

Table of Contents

Some Important Historical Dates and Events Relevant for this Study

1885:	The British Government declared the Oil Rivers the Protectorate of the Niger Delta.
1899:	The Royal Charter granted to the Royal Niger Company (RNC) withdrawn and transferred to the Niger Coast Protectorate.
1990:	The British Government `formally' colonized the Niger Delta.
1906:	Southern Protectorate and Colony of Lagos merged to form the Southern Nigeria.
1914:	Northern and Southern Nigeria amalgamated by Lord Lugard to form Nigeria.
1914:	The Colonial Mineral Ordinance enacted. This Ordinance vested all minerals mined in the colony in the Crown Government in London. It also prohibited non-British companies from operating in the colony.
1927:	Shell-BP exported petroleum products to Nigeria. Note: This was not the forerunner of the subsidiary of the Shell Group being studied.
1937:	Shell Petroleum Development Company began operation in the Niger Delta.
1938:	Exploration licence granted to Shell.
1951:	Shell Petroleum Development Company incorporated in Nigeria.
1951:	First Oil well drilled at Ihuo-north-west of Owerri (it was a dry hill, meaning no oil was struck).
1956:	First successful oil well drilled at Oloibiro.
1958:	The Mineral Act of 1914 amended.

1958: First crude oil exported to Europe by Shell at Port Harcourt.

1960: 1958 Mineral Ordinance amended and allowed non-British oil companies to operate in the country.

1960: Nigeria gained her political independence from Britain.

1961: Bonny Terminal Commissioned.

1961: Shell's office in Port Harcourt commissioned.

1965: First Petroleum Refinery commissioned by Shell-BP in Port Harcourt.

1967: Nigeria civil war began.

1967: The Nigerian Petroleum Decree enacted.

1968: Shell's loading terminal off Bonny coast bombed by the Biafran troops.

1968-70: Shell's operation in the Niger Delta suspended due to the civil war.

1969: The all-embracing Petroleum Decree No.51 enacted

1970: Nigerian civil war.

1970: Shell's first major oil spillage recorded due to the rupture of the Ejama/Ebubu - Bunny trunk line. Spilled for three weeks before it was stopped. All crude oil spilled was lost to the environment.

1971: Nigerian joined OPEC.

1971: Warri office of Shell was commissioned.

1971: Shell's Forcados terminal was commissioned.

1972: Nigerian National Oil Corporation formed.

1972: Lagos office of Shell (Headquarters) was commissioned.

1972: Nigerian Enterprises Promotion Decree enacted, repealed by the late Abacha junta in 1997.

1973: First participation agreement (joint venture) signed with Shell-BP. Federal Government 35 percent; Shell-BP 65 percent.

1974: Second participation agreement signed with Shell-BP. Federal Government increased its share to 55 percent; Shell-BP: 45 percent.

1974: Shell achieved a peak crude oil production of 1.4 million barrels per day.

1974: Petroleum Refining Regulation Decree was enacted.

1975: The Petroleum and Distribution (Anti-Sabotage) Decree was enacted.

1977: NNOC dissolved and NNPC established by Decree.

1978: The Land Use Act enacted.

1979: Associated Gas Re-Injection Act enacted.

1979: Third participation agreement signed with Shell-BP. Federal Government increased its shares to 60 percent; Shell-BP 40 percent.

1979: BP share in Shell-BP joint venture nationalized by Federal Government. Federal Government's share increased to 80 percent; and Shell's 20 percent.

1979: Shell-BP Joint Venture revoked in Nigeria.

1981: The Clean Nigerian Associates (CNA) formed by 9 foreign oil-producing companies and NNPC in Nigeria.

1986: Memorandum of Understanding between Shell and Federal Government signed.

1986: Introduction of Structural Adjustment Programme by the Babangida administration.

1989: Federal Government divested 20 percent of its participating share in the joint venture with Shell. Ratio: Federal Government 60%; Shell 30%; Elf 5%; Agip 5%.

List of Abbreviations

AWAM	Association of West African Merchants
CBN	Central Bank of Nigeria
CO	Colonial Office
CODESRIA	Council for the Development of Economic and Social Research in Africa
FO	Foreign Office
GRA	Government Reserved Area
IBRD	International Bank for Reconstruction and Development (World Bank)
IMF	International Monetary Fund
IPC	International Petroleum Company
JHSN	Journal of the Historical Society of Nigeria
NIIA	Nigerian Institute of International Affairs
NNOC	Nigerian National Oil Corporation
NNPC	Nigerian National Petroleum Corporation
NUPENG	National Union of Petroleum and Natural Gas (Workers)
OAU	Obafemi Awolowo University (Ile-Ife)
OPEC	Organization of Petroleum Exporting Countries
OSC	Oil Servicing Companies
OUP	Oxford University Press
PZ	Paterson Zochonis
R&D	Research and Development
ROAPE	Review of African Political Economy
SAP	Structural Adjustment Programme
SBM	Single Buoy Mooring
TNC	Transnational Corporations
UAC	United African Company
UDT	Underdevelopment Dependency Theory

USA	United States of America
UTC	United Trading Company
WWI	First World War
WWII	Second World War
£	The British Pound Sterling
US$	United States Dollars

Preface and Acknowledgement

As its primary objective, this study analyses in a holistic and dialectal perspective, the role of Shell Petroleum Development Company (Shell) and the state in the underdevelopment of the immediate host oil-producing communities in the Niger Delta of Nigeria. Taking Bayelsa, Delta and Rivers states as a case study, it critically examines the process and nature of the underdevelopment of these areas in particular and the Niger Delta generally, beginning from 1937 when Shell first began operation there, to 1997 when most of the contradictions arising from the relative reckless manner with which Shell has been exploring and producing crude oil in the region particularly the degradation of its environment, have fully unfolded. The study also analyses the pattern of reactions of the inhabitants of the oil-bearing areas to the state/Shell capitalist development there, noting the various actors/actresses, and institutions involved, and the interests they represent; the development of alternative strategies for survival by the people; and the responses of both the state and Shell.

This study is divided into nine chapters. The first three chapters identify and clarify the basic unsettled theoretical issues in the received literature on the political economy of the transnational corporations in their host countries in the Third World. Chapter One introduces the work and sets its scope and methodology. While Chapter Two analyses fundamental theoretical issues that underpin the political economy of the transnationals in the Third World countries. Chapter Three extends the frontiers of the discourse by taking a critical look at the nature of capitalist development that both foreign oil capital and the host state have promoted in Nigeria's petroleum industry and the oil-producing communities. All is done in an attempt to provide a rigorous theoretical basis for understanding how Shell has, in conjunction with the state, come to shape the contemporary economy of the company's host oil-producing areas in the Niger Delta.

Chapter Four analyses how British rule laid the foundation for the capitalist exploitation of the Niger Delta by Shell, among other foreign capital. Chapter Five then proceeds to examine the operation of Shell in the Niger Delta, taking cognizance of the antecedents of the major oil transnationals in the global petroleum industry. The next three chapters address the consequences of oil exploration and production on the host oil-bearing communities, noting the environmental degradation and rural poverty of these areas. In order to have a broad picture of the impact of the foreign oil companies operations on the Niger Delta, Chapter Six examines generally, the nature of the activities of both Shell, and other foreign oil companies like Mobil, Chevron and Texaco and their impact on the environment and economy of the host oil communities. This chapter also compares and contrasts generally, the activities of the foreign oil companies operating in the Niger Delta. As one of the major host oil-producing areas to Shell, Chapter Seven further examines how the company's pollution of the environment of the Ughelli oil communities has impacted negatively on the culture and traditional health system of the people who live there. Since land is most polluted by the oil companies, Chapter Eight critically looks into how Shell has, through its mode of production of crude oil, heightened land crisis in the Isoko oil areas. Chapter Nine concludes the study with policy prescriptions, charting a new course for the development of the Niger Delta.

First, I thank God, the Almighty, for seeing me through this project. I would also like to thank my wife, Josephine, and my daughter, Mudia, both of whom supported me both physically and spiritually in the course of writing this work. I owe a huge debt of gratitude to my elder brother, Jacob Omoweh, who denied himself virtually all the niceties of life in order that I can be educated.

In the course of this research, I received a lot of assistance from many quarters without whose help this work would not have seen the light of the day. I thank in particular, Julius Ihonvbere, Joy Ogwu, Amadu Sesay, Ogaba Oche, Adebayo Olukoshi, Rafiu Akindele, Kayode Soremekun, Isa Bukar, Bayo Kolade, Shola Ojo, Godwin Ikperite, Ezekiel Okpagoro, Isaac Maduikeh, Simeon Ugwu, Peter Eboh, among others. Many thanks also go to some officials of the NNPC, Shell, Mobil, Chevron, Department of Petroleum Resources and Oil Servicing Companies, who, having been disgusted with the recklessness of the oil companies, violated the "oath of secrecy" and at the risk of being fired, divulged classified information to me. Studying the activities of Shell in Nigeria was like daring the devil because of its connections across all sectors of the government of Nigeria. In fact, most of my field trips to the Niger Delta would have been fruitless, but for the cooperation that I received from the various youth organizations in the area. For this, I remain grateful. There are a lot more people who also supported me in one way or the other in the course of this work, but whose names I cannot mention here. I thank them all, believing that when they see the final product of this study, they would know that their effort was not in vain. I bear sole responsibility for whatever errors that this study might contain.

Daniel A. Omoweh
Nigerian Institute of International Affairs
Lagos, September 2004

Chapter 1

Methodological Context

1.1 Purpose of the Study

As its primary object, this study is concerned with the structure of underdevelopment engendered by the activities of Shell Petroleum Development Company (Shell) and the State in the immediate host oil-producing communities in the Niger Delta of Nigeria. Taking Bayelsa, Delta and Rivers States as a case study, it, in particular, examines the relative recklessness with which Shell still explores and produces crude oil in the Delta region, noting how the attendant environmental crises have, in turn, structured the emergent trend of underdevelopment of the inhabitants of the oil areas. In the past four and a half decades that Shell has produced and still produces crude oil in the Niger Delta, there was hardly any year it did not dump untreated harmful drilling wastes on land and into swamps, flare gas in both day and night, and spill crude oil without any clean up exercise and post-spillage environmental and socio-economic assessment. This is a level of recklessness which the company could not afford in the process of finding and wining oil in Europe and the USA. On its part, the Nigerian state, its politics and mode of surplus extraction made it possible for Shell to operate with reckless abandon. On account of the material interests of its custodian, the state collaborates with Shell among other oil transnationals (TNCs) operating the upstream sector of the oil and gas industry, to pillage the immediate host oil areas in the Niger Delta. Departing from the conspiracy theory of blaming the oil transnational corporations wholly for the underdevelopment of the Niger Delta, this study, therefore, examines how the state, together with Shell, degrade the environment of the host rural oil areas thereby deepening the incapacity of the people who live there to sustain themselves socio-economically.

Generally, copious works already exist in the literature on the role of the transnational corporations in the underdevelopment of the host countries most of which are Africa, Latin America and Asia. Raymond Vernon's *Sovereignty At Bay*[1], Greg Lanning *et al*'s *Africa Undermined*[2], and Clower *et al*'s *Growth Without Development* are among the notable works in the political economy of the TNCs in the Third World.[3] However, since the turn of the 1960s, the oil TNCs is particular, have, on the strength of the global high demand for petroleum as an energy source without any known substitute, been shaping the course of the development of the world economy. Yet, not much studies have really been done on the operations of the foreign oil companies in the developing economies, particularly in Nigeria. In Nigeria, what really exists in the literature on the subject, with the exception of few case studies, is an extension of the underdevelopment and dependency theory debate on the transnationals. Generally, the debates drew inspiration from theories of modernization but lacked concrete data to demonstrate the nature of underdevelopment. Worst still, some scholars were not regorous in applying Marxian methodology thereby leaving the meaning of underdevelopment either lost, or at best, vague and circulatory. So vague are the meanings of underdevelopment and dependency that most of the scholars depart from the mental construct of international bourgeois social sciences which they are out to criticize; and for which they intend to provide alternative theoretical framework for a deeper understanding of the activities of the transnationals in host countries. The issue at stake is not so much with the need for a new theory as many postulations abound on the subject. Rather, it is how to demonstrate the specificities of underdevelopment. So, this study does not reject the underdevelopment and dependency theory *per se*; nor the thesis that the transnational corporations are underdeveloping the host countries in the Third World. On the contrary, it contextualizes and concretizes underdevelopment; and, by so doing, seeks to bring out clearly, the nature of the underdevelopment of the Niger Delta as evidenced in the environmental crises, and socio-economic backwardness of the oil communities where Shell operates in the Niger Delta.

1.2 Methodology

In conducting this study, the basic methodological assumption made, which, in turn, structures the presentation and analysis of the empirical evidences of the underdevelopment of Nigeria's rural oil communities served by Shell and other foreign oil companies, is that of holism. Thus, in agreement with G. Lukac's position that,

> It was not enough to declare a political commitment for or against capitalism. One had to choose; either to regard the

> whole history of society from a Marxian point of view, that is, as a *totality*, and hence to come to grips with the phenomenon of imperialism in theory and practice; or else to evade this confrontation by confining oneself to the analysis of isolated aspects in one or the other of the specialized disciplines[4];

this study takes, therefore, the *Marxian method of historical inquiry* as its framework within which data are analyzed. The basic propositions of this body of theory are: (i) that society, which is always in constant motion cannot be understood outside the specific state of development of its production forces; and (ii) that no social, political and economic institutions can be understood independent of the mode of production within which they exist.[5] With the materialist conception of history, therefore, the development of phenomena in society will be understood only in their interconnectedness, and not in isolation in order to avoid static or empiricist analysis. The justification for employing the Marxian historical analysis lies in its capacity to combine the political, economic, social among other aspects of society in order to explain more rigorously, the historical past, and present situation of the oil-rich Niger Delta, and on the basis which credible projections are made for its future. In understanding the different stages of development in the Niger Delta, events are therefore, analyzed in a historical and chronological manner. Suffice it to note that, although Shell began operation in the Niger Delta in 1937, there is need to first examine the social formation of the area pre-dating the advent of the company in order to gain a deeper understanding of its underdevelopment since crude oil was discovered there. Such incursion into the history of the Niger Delta is very necessary since not only was it already in existence before Shell came in 1937, but had previously been invaded by Portuguese slave merchants in the late 19th century. All this will provide a deeper understanding of how the development continuum of the area was disrupted before Shell began oil operation there; and how its activities further disarticulated the economy of the area. In fact, it is only within the framework of Marxian historical inquiry that the present realities of the rural oil areas in Bayelsa, Delta and Rivers States of Nigeria can be properly studied. The crash of socialism in Eastern Europe since 1989, does not suggest that the utility of Marxism as a mode of analysis, is either rendered invalid, or buried with it. Rather, it is still the most rigorous methodological framework for analyzing the operations of Shell and the state in the Niger Delta because of its analytical strength.

In discussing the relationship between Shell and the state, the study does not simply analyse what they think of each other since such an approach will be too empiricist and over-generalizing. Instead, it looked into their evolution, noting elements of change being brought about by the changing circumstances in the society. For, it is through the historical materialist approach that a deeper understanding can be gained of how the pre-1937 events were altered since the advent of Shell, and the implications of the company's operations for the nature of the underdevelopment of the Niger Delta.

Furthermore, it is only with the adoption of the Marxian method that the discussion on the underdevelopment of Shell's host oil areas can be located in the context of the 'larger historical process', that is, the global process of capitalist accumulation, this time, in the petroleum industry of which Nigeria is an intricate part. Suffice it to note that, the underdevelopment of the Niger Delta by the forces of imperialism as represented by Shell, was part of the contradiction engendered by transnational capitalism in the global petroleum industry. Therefore, the study of the environmental degradation of the Niger Delta by Shell is one aspect of the paradox of the internationalization of Euro-American imperialism in the world oil and gas sector. Kwame Nkrumah summed up Africa's predicament as follows:

> Africa is a paradox which illustrates and highlights neo-colonialism. Her earth is rich yet the products that come from above and below her soil continue to enrich not Africans predominantly, but groups and individuals who operate Africa's impoverishment. If Africa's multiple resources were used in her development, they would place her among the modernized continents of the world. But her resources have been, and still are being used for the greater development of overseas interests.[6]

In Nigeria, the Niger Delta's experience brings into bolder relief, how Shell and the Nigeria state have not only furthered their financial interests from crude oil produced in the area, but also, enhanced surplus capitalist accumulation in the global oil and gas industry, leaving the host communities in particular, and the country at large, as 'the wretched of the earth'[7].

It is quite possible for a non-Marxist analysis to be materialist; however, the superiority of Marxist materialism which informs its adoption for this study, lies in its advantage in enabling us to come to grips with the conflicting interests between and among the social classes that make up the society and the struggle it gives rise to. For, it is these struggles within and among classes that dictate the nature of accumulation and direction of growth, development, and the underdevelopment of society not least the

Niger Delta. In examining the specific role of the Nigerian state in the impoverishment of its own people who live in the oil areas, very important attention is, therefore, paid to its historical evolution, noting in particular, its character and mode of surplus extraction in the country's petroleum industry.

Bourgeois social sciences have always regarded the state as an agent of development. Our contention in this study is not to refute that claim *per se*; for, the Nigerian state has been largely responsible for the inability of the country to reverse its underdevelopment. Instead, our concern is to deeply examine the nature of state capitalist development in the upstream sector of the petroleum industry, nothing the interests served, and its impact on the host oil areas.

It will be sheer hypocrisy to agree with Walter Rostow's thesis that Third World countries should follow the path that the West took to development in spite of the fact that most of the former were former colonies of the latter[8] So, also, will it be futile to rely on the postulations of main stream development theorists if greater understanding of the structure of underdevelopment of the oil-rich Niger Delta is to be gained. That is not all as the argument of some early UDT scholars that proper capitalist development in the periphery is 'blocked' has not even helped matters. Gunder A. Frank[9] might be right in arguing that rather than development, Third World countries have only experienced underdevelopment. But it is not too clear what he meant by underdevelopment and its structure. In what context therefore, will the underdevelopment of the Niger Delta be defined? These and other interrelated unsettled theoretical issues in the received literature are treated in details in Chapter Two.

In this study, I have tried to analyse underdevelopment as not just a sign of backwardness or stagnation, but a product of a specific development process that is closely connected with the rise of the world economy, and is the direct result of it. Therefore, it is only the analysis of the development of the capitalist world economy that can provide the key to an understanding of the present state and growth problems of the underdeveloped countries.[10]

Therefore, in discussing the underdevelopment of the host oil-bearing in the Niger Delta by Shell and the state, cognizance is taken of the nature and problems of monopoly capitalism in the global petroleum industry of which Nigeria's oil and gas sector is a part. For, it is with a deeper understanding of the dynamics of the world oil and gas industry that useful insights can be gained into how the forces of transnational capitalism together with the state, pillage the Niger Delta. Thus, *underdevelopment*, as used here, may bring about even appreciable increases in per capita income and provision of some basic social amenities in the Niger Delta, but only in a form that facilities the further exploitation of the oil areas which benefits Shell, the state and its allies and, it certainly imposes severe social and economic costs on the poor majority of the people whose primordial mode of production has been overthrown, and their traditional ways of life and overall internal capacities to reproduce themselves socio-economically have also been destroyed. *Development*, on the other hand, is not conceived here in the image of the western developed economies; nor does it mean the mere provision of some social amenities like water in a few oases in the oil areas whose larger environment remains deeply polluted. Instead, development, as used in this study, means the *desired* development of the people which involves a process of structural change and the accumulation of capital that enables the people living in the oil-producing areas in the Niger Delta in particular and Nigeria generally, to guarantee their basic human needs and to ensure an increasing social, cultural, political and economic equality, as well as a sustainable environment. At the heart of this development, is the improvement of the quality of life of the individual citizen.

Data in themselves, do not make much meaning; instead, but a lot depends on how they are obtained, interpreted and analyzed. Thus, while oral history, interviews and content analysis of documents are the basic methods used for data collection in this study, the Marxian political economy approach has been employed in the presentation of these data in order to synchronize the inter-connectedness between history, figures and reality on ground. The aim is to deepen understanding of the structure of underdevelopment of Shell's host oil areas in the Niger Delta.

The basic advantage of oral history, notwithstanding its shortcomings, lies in its capacity to provide first hand data, which are then used to either corroborate or refute existing works on the political economy of the transnational corporations and the state in Nigeria. No doubt, the interview method provides the respondents enough horizon to express themselves. It also allows the author much room to frame and reframe questions. But it is not without its shortcomings such as low response. However, they are not enough to jettison it. Data have also been obtained from library and archival sources. Because these sources are mainly books, journals and monographs, the method of extracting data from them is that of content analysis.

In attempting to provide a fuller understanding of the nature and process of the underdevelopment of Nigeria's oil-producing communities in the Niger Delta, this study first examines the primordial social formation of the country, inclusive of the delimited area, than the nature and process of its incorporation into the western capitalist system through British colonialism and finally, crude oil exploration and production by both Shell and the state, and the consequences for the host oil areas. A clearer picture of these and other related issues is presented under the period and scope of this work.

1.3 Period and Scope of the Study

This study covers a period of 60 years beginning from 1937 when Shell first arrived in the Niger Delta to explore and produce crude oil, up to 1997 when the long-term negative consequences of environmental pollution on the socio-economic well-being of the people who live in the host areas have fully unfolded. By 1937 when Shell came to Nigeria, the global capitalist petroleum industry was already dominated by 'mergers' formed by the giant oil TNCs. The anti-Trust laws in the American and Western European oil industry only forbade oil capitals from merging interests in their home countries. So, most of the mergers were formed when the oil transnationals began to move business to Africa, Middle East and Latin America. Why was the anti-Trust Law enacted in the first place? The United State of American enacted the Anti-Trust Law in 1911 to break the increasing monopoly power of The Rockefeller Oil Trust to which the growing American economy was enslaved. In essence, it was to save the expanding American industrial economy from excessive profit maximization by the oil giant. Also, the laws had ripple effects on the Western European's petroleum industry, especially as the major oil companies in that sphere like Shell Trading and Chemical Company and Royal Dutch Oil Company merged interests as they moved business to Africa and the Persian Gulf. Since the Anti-Trust Act was enacted, many small-equity based oil companies, otherwise known as the Oil Servicing Companies (OSCs). In fact, Shell Petroleum Development Company, the biggest oil transnational in Nigeria, is a merger of the Royal Dutch Oil Company of Holland and Shell Trading and Chemical Company of England.

No doubt, the formation of mergers by the oil capitals outside of Europe and the USA became noticeable after the Anti-Trust Act of 1911. The real problem however, was not so much with the Act as a response to the crises of over-concentration of capital, declining returns on investment and fast exhausting raw materials in Europe and America particularly since the end of the World War II in 1945. This was to inform the advent of Shell and other major oil companies like Mobil, Chevron, Texaco and Elf in Nigeria. From the onset therefore, Shell like any other oil transnationals, was not in Nigeria to develop the country. Instead, it can be argued that, the company only helped to reproduce the process of international capitalist accumulation in the global oil and gas industry. Some of the contradictions arising from the capitalist development process of the oil transnationals are technological dependence of the host country's oil and gas, and the degradation of the environment of the host oil areas. The internationalization of the underdevelopment of the Niger Delta was made possible through British colonialism drug which Shell first gained entry into the Nigeria's oil and gas industry. So, the significance of 1937 as the take-off date, lies in the early period of the deepening crises of accumulation that Euro-American oil capitals were faced with in the global petroleum industry.

As it shall be revealed later in Chapters 6, 7 and 8 later of this study, the ecological crises resulting from oil pollution take various times to manifest themselves. While it could take as short a period as two years to notice the incapacity of an oil-polluted soil to sustain the growth of crops etc., the cumulative impact of other forms of pollutions like gas flare, dumping of untreated drilling wastes on land and into swamps, and oil spillage, could take about a decade or more to be significantly noticed as in the case of a 'derived savannah.' Again, the overall unsustainability of the environment of the oil areas could be noticed after 15 years of oil/gas pollution; a trend that seems to dominate the oil areas since the turn of the 1980s - leaving the well-being of the people who reside there particularly worse-off. In terminating the period of the study in 1997, therefore, enough allowance was made for most of the negative consequences of oil exploration and production on the ecology and life of the people who live in the oil areas of the Niger Delta, to have fully manifested.

Taking the period 1937-1997, for this study, is by no means an assumption that it is static. Rather, the method of Marxian political inquiry demands that pre-1937 events have to be investigated in order to fully come to terms with the nature and trend of development in the Niger Delta pre-dating Shell. For, it is only by so doing that a better basis can be established to gain a fuller understanding of the extent to which the operations of the company together with the state, have impacted negatively on the pattern and direction of development of the Delta area since it came there in 1937.

In terms of scope, this study is essentially limited to the activities of Shell in the host oil-producing communities of the Niger Delta in general, and those in Bayelsa, Delta and Rivers states in particular. The concessional blocks of Shell in these areas are located in both *on-shore* (including swamps and land) and off-shore areas. Situated within Latitude 4°N - 6°N of the Equator, and Longitude 4°E - 8°E of the Greenwich Meridian, they span through the following states in the Southern tip of Nigeria: Akwa Ibom, Anambra, Bayelsa, Delta, Edo, Imo, Abia, Ondo and Rivers. Since Ondo state was split into two states namely Ekiti and Ondo, in October 1996, Shell's skeletal oil activities are now in the latter. In all, this study is restricted to Bayelsa, Delta and Rivers states where about 90 percent of the country's oil is currently produced; and where most of the company's operations with the attendant ecological crises are equally concentrated.

Limiting the geographical area of study to the three states does not carry the assumption that the rest six states are not worthy of study. Rather, two factors were taken into account in delimiting the territorial area for this study: the scale of operations and the impact of the activities of the company on the host areas. Shell has skeletal operation around Ilaje area of the present day Ondo state which shares a common border with Edo state. The company's operations around the Onitsha angle of Anambra state are also scanty. They are, therefore, too small to merit a study of this magnitude. In both areas, Shell can be described as still operating marginally, and is on the verge of pulling out as the oil wells therein are becoming extinct.

The initial attempts to explore for crude oil by Shell in Nigeria began around Owerri in the present Imo state, particularly the Akata I and II oil-wells which were drilled in 1937, but with no commercial quantity of crude oil therein. With further exploratory activities, the company found oil around the Egbema/Oguta communities which share boundary with Rivers state. Shell's level of operations was and is still low in Imo state when compared with those in Bayelsa, Delta and Rivers states. In Akwa Ibom state, it was and still remains minimal. Given the skeletal activities of the company in these areas, therefore, they hardly merit a study of this degree. However, references are still made to them in order to clarify or refute arguments.

Shell's operations are both off- and on-shore. This study looks into its operations in the on-shore areas; for, the off-shore locations are not only few but are also usually sited in large water bodies where the economic activities of the people like farming and fishing are minimal. That is not all as it is really difficult to assess the impact of Shell's ecological pollution in the off-shore areas because of sea waves as evidenced in the incident of one of Shell's sunk oil rig in the Bonny area in 1985. Being off-shore, it was difficult to estimate the barrels of crude oil lost and effects on the ecology. However, references were made to such cases to enrich our understanding of the company's relative recklessness in Nigeria's Niger Delta.

Oil production is a very specialized activity. Beginning from exploration to the point where the crude oil gets to the terminals ready for export, many specialized companies are involved. Of particular significance, are the OSCs such as Schlumbeger, Flopetrol, which as explained earlier, are bye-products of the US government: Anti-Trust Law in 1911. Because the servicing companies are very specialized oil companies and of small equity base, in order to survive, they have to carry out the bulk of the jobs of finding, producing and transporting crude oil among other services in the upstream sectors for the major oil transnationals like Shell, Mobil, Texaco, Chevron. Since the OSCs explore and produce the crude oil, Though it is the OSCs like Dowel, Flopetrol and Otis they really degrade the environment of the oil areas. But they do not constitute a separate focus of study. Since they are contractor companies and constitute part of Shell's capitalist mode of operation. Lastly, oil production entails a lot of technicalities. This study only makes references to them in so far as they helped clarify some basic issues raised in the discussion.

Chapter Two seeks to deepen the theoretical understanding of the nature of underdevelopment of the Niger Delta which was brought about by the operations of Shell. It begins by tracing the origins of the debate on the transnational corporations and the underdevelopment of the host countries in the Third World. Thereafter, it situates the discussion proper specifically in the context of the political economy of Nigeria in order to bring into greater light, the nature of the alliance between transnational capital, the state and its local allies in the petroleum industry, noting its implications for the material well-being and environment of the people in the oil areas. As an extension of the theoretical issues raised in the previous Chapter, the main task of Chapter Three is to enhance our understanding of the political economy of underdevelopment of Nigeria's petroleum industry and the Niger Delta.

Chapter Two focuses on how the foundation for the underdevelopment of Nigeria including its oil areas was laid by British colonialism. In doing this, it examines the pre-capitalist period of the country's economy in order to gain a deeper understanding of how the imperialist forces not only cashed in on colonialism to further their selfish economic interests but set the stage for Shell's exploit and that of other oil transnationals operating in the Niger Delta.

Thereafter, Chapter Five addresses, in detail, the advent of Shell in Nigeria. It begins with the history of the company and moves on to locate its operations in the global petroleum industry to help unfold the origins of the contradictions engendered by the that monopoly capitalism in the global oil and gas industry of which Shell is an intricate, as well as its part, implications for Nigeria's Niger Delta. Lastly, it critically looks at the structure and mode of operation of the company in the country, noting its impact on the environment of the host oil areas. That Shell is the chief culprit in the degradation of the environment of its host communities does not mean other operating oil companies like Mobil, Chevron (Gulf), Texaco, Elf, Agip, and Philip are not polluting. As its major aim, therefore, Chapter Six first presents the picture of the general trend of environmental pollution arising from the operation of Shell, Mobil, Chevron, Texaco etc. and resultant socio-economic consequences across the oil areas of the Niger Delta. It, in particular, examines the nature of rural poverty in the host oil communities, analyses the pattern of protest across the areas, noting the major actors and actresses, and institutions involved and the interest they represent; and the responses from both the oil companies and the state.

To further unearth the extent of the underdevelopment of Shell's host oil communities, Chapter Seven takes a case study of the Ughelli oil communities. It examines in detail, the magnitude of destruction done to the sociological basis of the culture of the people particularly their indigenous ways of life, and traditional health system of the people. It further analyses the reactions of the people to the activities of Shell in the community, and the responses from both the company and the state.

Chapter Eight examines the implications of Shell's operations for land crisis in the rural Isoko oil-producing areas, noting the resultant effects on the social and economic welfare of the people. It critically looks into the emerging strategies adopted by the people to cope with landlessness and rural poverty, the form that protests have taken in the area, and finally, the responses from Shell and the state. This study is geared towards ensuring a sustainable environment that will permit the desired development of the oil areas to take place. It is this task that Chapter Nine, the Conclusion, sets out to achieve. It first summarizes the findings of this study then examines their implications for theories on underdevelopment, and finally, recommends policy prescriptions.

1.4 Case Study on Shell and the Immediate Host Oil Communities: Its Significance

This study lays claim to originality, a claim that stems from its demonstration of the concrete structure of underdevelopment of host oil areas by the oil transnational corporations operating in Nigeria's petroleum industry. Apart from Adebayo Olukoshi's[11] study which unfolded the nature of industrialization that the transnationals foisted on Northern Nigeria with specific reference to Kano, the exact impact of their activities on the country's economy has been plagued largely with over-generalization as evident in the received literature in the field. Bade Onimode[12] and Okwudiba Nnoli[13] have both presented surveys on the activities of transnational corporations in Nigeria, arguing in their respective works that, the transnationals are indeed, the major cause of Nigeria's economic underdevelopment. To them, the profit motives of the TNCs and rudimentary skills which they bring to Nigeria in the name of technology, do not conform with the developmental goals of the country. The issue at stake is not whether they are right or wrong. Rather, it is the empirical evidences that really informed their conclusions, which both studies could not readily produce. In the petroleum industry, there is hardly any concrete study on underdevelopment of the oil areas. It is not to claim that studies have not been done on the oil sector. Far from it. Instead, a lot of the studies were based on the linkage between oil and Nigeria's foreign policy. In fact, very little is done in these studies to deepen the understanding of the political economy of the oil TNCs in Nigeria, much more their role in the underdevelopment of the Niger Delta. Through the operating foreign oil companies havoc was wreaked on the environment of the host oil communities with the consent of the state, no case study on any of them has been done. Except for this study on Shell, the most powerful oil TNC controlling more than half of the activities in the country's oil sector, a lot of the existing works on the oil TNCs in Nigeria are general. Yet, there is the general claim among scholars, activists, policy makers that they (oil TNCs) have underdeveloped their host areas. The question is: what is the basis of such claim? Until the eruption of the Ogoni case in 1993, there was no major research interest was demonstrated on the oil areas. Even since the Ogoni's uprising, the emerging literature on the topic focuses so much on minority politics that the structure of underdevelopment of the area is not properly contextualized.

The study is, therefore, a ground-breaking effort towards filling such gaps in the literature. It departs from the previous works, being the first time a deep examination of the environmental degradation and the rural poverty of the oil areas caused by the operations of Shell, largest oil transnational in Nigeria, is undertaken.[14] Furthermore, it focuses on Bayelsa, Delta and Rivers states, three states which have been particularly worst hit by environmental pollution and protests. To that extent, this study can be regarded as being a pioneer attempt to concretize the structure of underdevelopment of Shell's immediate host oil communities. The need to focus on the environmental underdevelopment of these areas is rooted in the fact that, all social and economic activities of the people in the area, and any other area for that matter, are tied to it. Invariably, degrading the environment causes widespread worsening of rural poverty among the people as they, at the time of the study, had neither fertile land to farm, nor rivers and creeks to fish.

Oloibiri, the first host oil community to Shell where it struck crude oil in 1956, was left wretched and wiped out of the company's map of flowlines following the extinction of its well after 35 years of rapacious exploitation. Oloibiri's experience is a pointer to the trend of the nature of the underdevelopment to expect in other oil areas in the Niger Delta[15]. The ultimate issue being raised here is that the nature of backwardness which the oil TNCs engender in their host countries, as in the case of Shell in Nigeria's Niger Delta calls for a rigorous theoretical rethinking on underdevelopment. It is the task of sharpening our theoretical understanding of the nature of the underdevelopment of the Niger Delta by the operations of Shell and the state, that Chapter Two is set to achieve.

Notes and References

1. R. Vernom, *Sovereignty at Bay: The Multinational Spread of US Enterprises*, New York: Basic Books, 1971.
2. G. Lanning and M. Mueller, *Africa Undermined: Mining Companies and the Underdevelopment of Africa*, Harmondsworth: Penguin Books, 1979.
3. R.W. Clower, G. Dalton, M. Horwitz and A. Walters, *Growth Without Development*, Northwestern University Press, Evanston, 1966; and J. Gus Liebenow, *Liberia: The Evolution of Privilege*, Cornell University Press, 1969.
4. G. Lukacs, 'The Marxism of Rosa Luxemburg' in *History and Class Consciousness*, London: Merlin Press, 1971, p.30.

5. K. Marx, *A Contribution to The Critique of Political Economy*, Moscow: Progress Publishers, 1970.
6. K. Nkrumah, *Neo-Colonialism: The Last Stage of Imperialism*, New York: International Publishers, 1965, pp.1-2.
7. F. Fanon, The Wretched of the Earth, New York: Groove Press, 1963.
8. Walter Rostow, *The Stages of Economic Growth*, New York: Cambridge, 1960, and his edited volume, *The Economies of Take-Off Into Sustained Growth*, New York: St. Martins, 1963.
9. Gander A. Frank, *Capitalism and Underdevelopment in Latin America: Historical Studies of Chile and Brazil*, New York: Monthly Review Press, 1967.
10. T. Szentes, *The Causes, Criteria and International Aspect of So-Called Economic Development*, MTA, Budapest, 1969, p.2; See also, S. Katz, *Marxism, Africa and Social Class: A Critique of Theories*, Montreal, McGill University, 1980; A. Foster-Carter, 'From Rostow to Gunder Frank: Conflicting Paradigms in the Analysis of Underdevelopment', *World Development*, Vol.4, No.3, March 1976; and E. Brett, *Colonialism and Underdevelopment in East Africa: The Politics of Economic Change 1919-1939*, New York, Nok, 1973.
11. Adebayo Olukoshi, 'The Multinational Corporation and Industrialization in Nigeria: A Case Study of Kano, C. 1903-1985, Unpublished Ph.D. Thesis, Leeds University, 1986.
12. Bade Onimode *et al*, *Multinational Corporation in Nigeria*, Ibadan: Shraden, 1983.
13. Okwudiba Nnoli (ed), *The Path to Nigeria's Development*, Dakar: CODESRIA, 1982.
14. Shell accounts for about 60 percent of the country's total yearly crude oil production estimated at 500 million barrels.
15. Daniel Omoweh, *The Political Economy of Shell Petroleum Development Company in the Oloibiri Areas of Nigeria*, (Stuttgarts: Bread for the World, 1998).

Chapter 2

Transnational Corporations and Underdevelopment of Developing Economies: A Theoretical Perspective

2.1. Introduction

With the replacement of coal by petroleum as the world's cheapest and leading source of energy with yet no known substitute since the end of the World War II in 1945, the oil transnational corporations have emerged as very influential non-state actors shaping not only the nature and direction of the economic and environmental underdevelopment of the host oil-producing and exporting developing countries usually found in Africa, Middle East/Asia and Latin America, but also, the political economy of development of the oil-consuming countries in the North. Particularly badly hit by the policies and nature of operations of the giant oil companies, is the environment of the immediate oil-bearing communities of the host countries; communities that have been faced with serious and frequent pollution of their environment since oil was found there so much as that they are increasingly becoming incapable of regenerating themselves. In fact, the material well-being of the people who live in the immediate oil-producing areas has, on balance, been worse than what it was before the oil companies began operation there.

Yet, from the received literature on the transnationals particularly on the activities in the host countries scholars differ as to what impact the foreign companies have had on the economy of the host states[1]. Two broad schools of thought have emerged on the debate on the political economy of the transnationals in the developing economies. First, is a group of scholars who argue that, the transnationals inclusive of those in the oil industry, have actually brought about significant economic development of the host-countries citing as benefits to the host countries, rents accruing from royalties, taxes and other forms of fees paid by the operating foreign oil companies. Also regarded as benefits to the host communities, are social amenities like (access) roads tarred which they construct to facilitate their operations. And that rent accruing from the oil companies constitutes a major source of revenue to the government of host states with which they embark on economic and social development. The position of this school is well taken. However, certain basic issues are still unsettled. What kind of development do the transnationals promote in their host countries? Whose interest does such development really serves? It is in an attempt to address these unsettled issues that gave rise to the second group of scholars.

Taking the (oil) transnationals as really no less than forces of monopoly capitalism, the second group of scholars contend that, not only does their profit motive run counter to the development goals of the host states, but helps reproduce the increasing incapacity of the host economy to sustain itself socio-economically. And that it is in the contradictions which inhere in the capitalist development promoted by the oil companies that the root causes and nature of the underdevelopment of the host states is to be located.[2] Contradictions that stem from the nature of the collaboration between the forces of state and transnational capitalism in the oil and gas industry.

In the next two chapters, therefore, our major goal is not to restate the propositions earlier advanced by scholars in the two broad schools of thought on the political economy of the transnationals *per se*. On the contrary, it is to deepen understanding of the main unsettled theoretical issues political economy of underdevelopment of the immediate oil-bearing communities in Nigeria's Niger Delta. For, it is with a fuller knowledge of these theoretical issues that a clearer picture of the specificity of the underdevelopment of the Niger Delta can actually be gained. In dividing this section of the study into two chapters, at a chronological presentation and therefore, the primary aim is to cease on understanding of the theoretical issues that underpin on the underdevelopment of the Niger Delta.

Thus, this chapter begins by tracing the origins of the basic theoretical issues that have come to inform the early debate on the political economy of transnational corporations in the developing economies. In collapsing the contributions to the debate into two broad groups namely, *the modernization theorists* or *development school*; and the *underdevelopment and dependency theorists or dependencia school*, it critically examines and analyses the opposing views of these scholars with the hope of bringing into bolder relief, the nature of state-transnational capitalism in Nigeria's economy inclusion of the oil industry, noting how the contradictions it gave rise to, in turn, structure the country's economic underdevelopment particularly that of the Niger Delta.

Chapter Three extends the frontiers of the theoretical discourse on the political economy of the transnationals to Nigeria's petroleum industry and the host oil areas in the Niger Delta. It examines and analyses the basic theoretical issues underlining the nature and interests of the major social forces namely, the state and foreign capitals which operate the country's petroleum industry, noting the consequences of their operations for the environment of the host communities in the Delta area. Let us now begin the discussion proper with the analysis of the debate on the transnationals.

2.2 Transnational Corporations and Underdevelopment of their Host Countries: A Debate Revisited

Scholars writing on the activities of transnational corporations in Third World countries are yet to agree on the precise role that they play in the underdevelopment of their host countries. The discourse on the transnationals can indeed, be collapsed into two broad but not absolute schools of thought: (i) *the development school*, and (ii) the *dependencia*. Let us begin the discussion with the development school.

2.2.1. The Development School

The early origins of the main issues that informed the theoretical position of scholars of the development school on the transnational corporations can be traced to the neo-classical theories on economic development. John M. Keynes, who, was basically concerned with the problems of depression and high unemployment in the early period of industrialization in Europe, particularly its capacity to solve them, provided the theoretical foundation from which this school took off. According to Keynes, an increase in expenditure, increases aggregate demand which would eventually lead to an increase in the level of economic activity and a decrease in unemployment.[3] No doubt, a depressed economy needs stabilization if economic development is to take place as Keynes has rightly suggested. However, Keynes was so pre-occupied with the short-term problems in the economy that he played down the long-term problems. For instance, he was indifferent to the problem of chronic capital shortage in the economy which is not only long-term in nature, but also central to its survival. This and other related long-term economic problems that Keynes failed to address were to be taken up by two notable economists, Evsey Domar and Roy Harrod[4], who came after him. Both Domar and Harrod built on the long-term problems of economic development, particularly since there is an intricate relationship between an economy's rate of growth and capacity to develop, and its level of savings and investment. The *Harrod-Domar Model*[5], named after the two economists, came to dominate the discourse on theories of economic development in the early 1950s. The Model took both growth and development as synonymous. So, too, were other development economists of their times who came after them. This was in spite of the fact that they are nothing of the sort. It is no surprise therefore, that in most of the literature on the field that time, no distinction was made between underdeveloped and undeveloped countries; nor between industrialized and pre-capitalist economies. What is more, the problem of the underdevelopment of former colonies was conceived mainly as that of shortage of capital.

As economic development became the major agenda of governments across Europe in the 1950s, development economists took centre-stage in economic planning, propagating ideas on how it can be achieved and sustained. W.W. Rostow[6], one of the notable economic development theorists at the time, made an in-road into the debate. He prescribed five stages through which all societies, developed and developing, have to pass through in order to reach a self-sustaining economic growth. According to him, *first*, is the *traditional stage*; a stage at which the level of technological knowledge is so low and imposing an upper limit on *per capita* production. The economic prerequisites that will enable societies move out of the first stage are created during the *second stage*, a stage where the characteristics of the traditional society must have been removed. Agricultural productivity at this stage, would have increased so rapidly; so, too, is infrastructural development. All this will usher in the *third stage*, regarded as the 'take-off' period which is very crucial to the furtherance of economic development. At the *fourth stage*, most of the obstacles to economic development would have been removed. The most significant aspect of the stage, is the expectation that the share of net investment and saving in national income would have risen from 5 to 10 percent. Others are that, modern technology would have been so largely disseminated from the leading sectors of the economy, and the economy would have been put on the path to maturity. At the *final and fifth* stage, the society would have attained mass consumption, the basic needs of the people met, and consumption shifted towards durable goods and services.[7]

The key element in Rostow's doctrine is the process of capital formation, which has been different in both the countries of the North and South. Colonialism made his five stages of economic development hardly relevant for countries that were previously under foreign rule. For, they had their development continuum first disrupted and later re-oriented by forces of imperialism to cater for metropolitan corporate capitals during the period of European colonialism. A process of disarticulation that continued into the post-colonial period.

While Rostow was pre-occupied with the various economic stages that societies should pass through in order to attain economic development, other scholars like David Richardo was interested in how to overcome the problem of shortage of capital. In advancing the theory of comparative cost therefore, Richardo argued that, a country may raise its level of consumption above what would have been possible in a state of autarky, but specializing in the production and foreign sale of commodities which, comparatively, have the lowest cost of production. The world as whole, as well as each single country (including the less developed ones) would obtain more goods through an international division of labour in which all nations specialize and export those commodities which they can produce locally[8], and in the process, raise capital with which to embark upon economic development.

Like Richardo, Arthur Lewis was concerned with how to raise the necessary capital for economic development. In formulating the 'dualist mode', Lewis basic argument was that, the capitalist sector would draw endlessly on the large underemployed labour reserves in the subsistence sector, a process which will free land in the latter sector for more food production and promotion of capital accumulation in the modern sector. He insisted that since the capitalist sector in the subsistence sector is still weak, foreign capital should be invited, and where it already existed, strengthened[9]. In fact, this formed the basis of his treatise on 'industrialization by invitation' for the developing economies.

From Keynes through Harrod and Domar, to Rostow and finally, Lewis, it is quite obvious that the early origins of the theoretical discourse on economic development was rooted in the historical experiences of Western Europe. These theories are not useless per se. Rather, the problem resides with their applictions in Third World countries. This is all the more so since they were creations of foreign capital; and were never part of the North; nor had common historical past.

Yet, studies on the economic development of Third World countries showed that their planning was based on the Western economic development models. Model that were applied on the general assumption that since they worked well for the West, nothing should stop them from being effective in the Third World.

For instance, the application of these models was premised on the assumption that for economic development to take place, there has to be transformation from a state of rural and agricultural underdevelopment to that of dominance of society by the urban and industrial sector, a process which is taken as a given and without the political underpinnings. It was also assumed that, with the provision of adequate capital, underdevelopment will be solved, believing that foreign capital will invest so freely by using appropriate technology, and bringing in liquid capital, instead of raising funds internally. And that free trade, conceived in absolute terms, will flourish in the international capitalist system in spite of the exploitative relationship with which it ties the Third World to the North. There was also the believe among the mainstream development economists that labour will be released freely without any resistance; and that development, understood solely in the context of economics, is synonymous with growth.

Nothing can be more misleading than these wild assumptions. For, there is nothing which suggests that the pre-industrial period of Europe resembles that of the underdeveloped countries of Africa. Nor was the historical past of Europe in any way determined by the same social and political forces tht helped in recreating Africa. While European colonial capitalism structured the economy of former colonies in Africa, the same cannot be argued for Europe. All this partly explains why efforts of the African countries at industrialization are not bringing about the desired results, but only further re-integrating their economies into the global capitalist economies. So, too, it all largely explains why Shell, as foreign capital, has precipitated the environmental underdevelopment of the Niger Delta in the process of exploring and producing crude oil in the area.

Yet, most of the early scholars who employed the development/modernization perspective to the studies on the transnationals in the Third World argued, that they, TNCs are indeed, the real hope for the economic development of the host states in the developing economies. This was all the more so as the general contention then was that, the host countries in the Third World had no viable economies and lacked critical production inputs like skilled manpower, capital and technology, which only the TNCs could provide. In essence, it was only the TNCs that could provide what was regarded as the 'missing links' in the scheme of economic development, a position still being held by the development school. Notable among the early scholars who advanced this position, was Peter Drucker. Drawing on his study of Latin American, Drucker argued that, prior to the advent of the TNCs, the sub-region's extractive, agriculture, manufacturing, finance and public utilities sectors were not developed. But with the arrival of Anacoda Copper for minerals; United Brands for agriculture; Chase Manhattan Bank for Finance; Ford Motors Company for manufacturing; Hilton Hotels for service; and ITT for telephone, the Latin American economy witnessed a dramatic development in terms of the exploitation of natural resources, provision of social amenities, jobs creation and industrialization. To him, therefore, the TNCs are the 'real hope' for the development of Latin America.[10]

Drucker was not alone in reaching such wild conclusion. In their combined study of the operations of the TNCs in Latin America, both David Blake and Robert Walter, too, had demonstrated that, the transnationals were more effective in exporting their products, especially manufactured ones than the domestic firms; and that, since they had the technology that gave birth to new products, there was the tendency for Latin American industries to acquire the technology that would enable them increase their competitive advantage based on low labour costs.[11]

Subsequent studies on the operations of TNCs in the Third World such as the one undertaken by Richard Barnet and Ronald Muller also alluded to the same sweeping claims made in the previous studies, that is, the transnationals as the 'real hope'. Taking the Third World as scope of their study, Barnet and Muller examined the activities of selected 500 United States transnational corporations to ascertain the impact of their operations on the economies of the host states, and that at the United States as well. They demonstrated that the major concern of the American TNCs operating in the Third World was basically to exploit its vital raw materials, utilize its abundant and cheap labour force, and take advantage of its large population to market their products. Their study also identified three main institutional weaknesses why Latin American countries (with the exception of Argentina and Honduras) and those of Africa for instance, could not effectively bargain with transnational capital. They were: (i) the inability of the host states to enforce their laws on foreign investment citing the case of Liberia where the United State government even wrote the foreign investment code of the country; (ii) lack of strong labour movement and union that can negotiate effectively with the TNCs; and (iii) lack of local businessmen to compete with them.

Finally, the study showed that Third World countries willingly sacrificed their build-up for technological capacity and finance capital by imbibing the western philosophy of development, pointing to basic social problems of these countries, as mainly those of poverty, unemployment and inequality, which, the operations of the TNCs only helped exacerbate in the 1960s. The situation has worsened since then.

On the implications of the operations of the TNCs for the economy of the United States, Barnet and Muller also demonstrated that, in transforming the world economy into a global shopping centre and factory, the main interest of the TNCs was the global maximization of profits; that the TNCs were being managed in such a way that a global system in which the long-term role assigned to the US itself, was completely changing what it produced and consequently what its people would do; and that as the cars, television sets, computers, cameras, clothes and furniture were being produced abroad, the US economy became a service one which produced plans and programmes for the Third World to execute.[13] This has not really changed.

No doubt, both Barnet and Muller raised critical theoretical issues that have come to interest the dependency writers. For instance, they portrayed the TNCs as agents of underdevelopment. Yet, as scholars, their theoretical disposition was neo-conventional. This was clearly evidenced in their basic position that, "if the TNCs had underdeveloped the host countries in the Third World, why were they still welcomed?" Suffice it to note that in spite of their critical appraisal of the TNCs, like Drucker, Blake and Walter, they also regarded the TNCs as agents of development in the host states of Third World. Not even Raymond Vernon's harsh criticisms against the TNCs made him any different from Muller *et al.* It is therefore, misleading to take seriously, Vernon's argument that, rather than the managers of the TNCs seeing the international system as made up of nation-states with all the anti-social measures that accompany the struggle for power, they operate to undermine the state system and contribute to the welfare of mankind by seeing the world as one. That is, the establishment of a common regional culture and life style.[14] For, Vernon still regards the TNCs as agents of change and development, altering traditional value system, social attitude, behaviour pattern, and encouraging responsibility among the political leadership of the less developed countries.

The real issue at stake is not whether the TNCs bring about development in the host-state, but its kind and the ultimate interest served. In fact, an insight into the nature of their operations and why they shifted business to the Third World, will shed light on their impact on the host states. Significant to note, too, is the historical past of the host states, particularly the role that colonial capitalism had played in shaping the nature of the post-colonial economy, and that of imperialist forces in re-inserting the neo-colonial economy into the world capitalist system. These critical issues were neglected in the studies by Drucker, Blake, Walter, Barnet, Muller and Vernon. They ought not to have overlooked these issues particularly at a time when the problem of over-concentration and centralization of international finance capital had intensified in Europe and the United States of America since the end of World War II in 1945. They would have equally taken note of the fact that it was also a period when international finance capital, a merger between monopoly banking and industrial capital, represented by Shell in the case of this study, has been seeking for sources of cheap labour, raw materials and profitable outlets for surplus capital outside of Europe and North America. These issues are too critical to be ignored as they did in their studies if a deeper understanding of the political economy of the TNCs in the Third World is to be gained. It is for this concern that an alternative methodology was sought to enrich the understanding of the operations of the TNCs in the developing economies. And no other group of scholars did this other than those employing the underdevelopment and dependency paradigm. It is to this, that we now turn.

2.2.2. The Dependencia School

Unlike scholars of the development School, those who employing the dependency paradigm in their studies on the political economy of the transnationals in the developing economies, can be regrouped into the structuralists, and the underdevelopment and dependency theorists. Notwithstanding the slight distinction in their methodology, both group of scholars, are unanimous in their conception of the TNCs as no less than expanded colonial trading companies, purveyors of over-priced obsolete equipment and technology, rapacious exploiters of labour and raw materials, major agents of profit repatriation and capital and resource outflow the rigour in their frameworks differs. However, while the structuralists stopped at identifying and describing the indices of underdevelopment, the scholars employing the UDT were concerned with the dialectics of underdevelopment. For instance, while the UDT adopts Marxian methodology, the structuralists are basically descriptive in approach. More of these distinctions will unfold in the course of the discussion.

2.2.2.1. The Structuralists

That there is a consensus among the dependency scholars that the transnationals underdevelop their host countries in the Third World, does not mean that they all employed the Marxian method of inquiry in their studies. It is by no means an assumption that only the Marxian methodology can be employed in the study of the TNCs in the developing world. The structuralists, who used non-Marxian methodology in their studies have also showed that the TNCs are agents of underdevelopment. What is lacking in their mode of analysis is the absence of the dialectics of history, the very strength of the Marxian method. Suffice it to note that, the structuralist mode of analysis is a logical extension of the challenges of the neo-classical economic theories on economic development.

Early Latin American development economists who were working with the Economic Commission for Latin America (ECLA) in the 1950s notably Raul Prebisch laid the theoretical foundation for this school of thought. As then Executive Secretary of the ECLA in the 1950s and 1960s, Prebisch was worried about the structure of the Latin American economy, particularly its disadvantaged position in the lopsided international division of labour, in which the sub-region was assigned the role of supplying raw materials to, and importing finished goods from the industrialized countries, especially the United States of America. He was equally concerned with the chronic balance of payments problems that Latin American countries were faced with since they came in contact with the forces of transnationals capitalism through the aegis of European colonial rule; and the predominance of the US TNCs in the region's economy.

In his study on the problems of economic development of Latin America, therefore, Raul Prebisch did not only demonstrate the structural crises the sub-region was faced with, namely those of balance of payment, primary producer and exploitation, but also, other features of underdevelopment like disarticulated extractive industry industries enclave operations, having little or no direct relation to the local economies. Prebisch's study also showed that, the foreign enterprises repatriated exorbitant profits to their home countries thereby causing serious drain on the balance of payments of the countries in the sub-region.[15]

Like Prebisch, Celso Furtado, another leading Latin American development economist of the structuralist genre, had demonstrated in his study that, although industrialization was initiated in Latin America by the transnationals, it did not automatically continue by itself. In the case of Brazil, a country he, Furtado, among other development economists, believed was already in the Rostovian take-off stage', his study showed that the country's industrial growth not only was conditioned by the TNCs, but also, that the Brazilian local bourgeoisis was totally devoid of initiative and drive, wholly dependent on government and foreign capital, and incapable of fulfilling its historical mission of raising the productive powers necessary for the desired development of the people.[16]

Drawing on the experiences of Chile, another major host country to the TNCs in the sub-region, Osvaldo Sunkel demonstrated in his study how the country's economy was pillaged and brutally undermined by foreign capital.

Prebisch, Furtado, Cardoso and Sunkel identified one common problem with the economy of Latin America: stagnating and increasing dependence on the United States' TNCs. While these scholars might be right in claiming that the transnationals were the root cause of the stagnation of the economy of Latin America, they did not provide a rigorous basis for understanding its underdevelopment. As colonies of European imperial powers, the nature of the post-colonial economy of Latin America has its origin in colonial capitalism. So, too, is the nature of the state in post-colonial Latin America. Yet, Prebisch *et al* failed to explain the dialectics of history, and the politics of the state in Latin America in the period since political independence was attained. Their inability to capture these crucial issues, resulted in the recommendation of wrong remedies. Which is why the adoption of import substitution industrialization (ISI) as a potent antidote to reversing the sub-region's underdevelopment, ended deepening its integration into the world capitalist system. In fact, the experiences of Latin America with the ISI under the aegis of the TNCs showed that, rather enhance,it deepened its incapacity to produce locally, hitherto imported goods. The ultimate result is the import reproduction instead of import substitution of previously imported goods. Suffice it to note that, it is enough to describe Latin America's dependence on the US TNCs, and its overall underdevelopment as Probisch *et al* did.

Therefore, the early origins of the dependency school proper can be traced to the contradictions inherent in the propositions advanced by the structuralists in their studies. Contradictions that eventually gave rise to a group of scholars whose main concern was to address the origins and dialectics of the underdevelopment and dependence of Latin America and other peripheral social formations. But before going into a detailed discussion of the basic propositions of this group of scholars, it is important to note that, since Latin America had about the greatest concentration of United States and many Western European based transnationals among other developed countries in the 1950s and 1960s, most of the early studies on the TNCs in the Third World were drawn largely from the experiences of the sub-region. This explains in part why most of the analyses in this study also from studies on Latin America. It is not to assume, however, that studies have not been done on the operations of the TNCs in Africa; rather, they are just beginning to emerge. Now back to the discussion proper.

2.2.2.2. Underdevelopment and Dependency Theorists

To begin with, there are as many theorists on the underdevelopment and dependence of the Third World as there are varied propositions. Of importance to us in this study, however, are the common crucial issues in the received literature that underline the political economy of underdevelopment and dependence of social formations in the Third World. Issues like the state and its mode of surplus extraction, foreign capital, local private capital, and domestic bourgeoisie.

One of the most influential structuralist who to first radicalize the ECLA analysis of why Latin America was underdeveloped, was Celso Furtado. As Director of Planning of the poorest regions of Brazil, he personally experienced the development of events that led to the military take-over of the Brazilian government in 1964. To him, lack of capital was not so much the cause of the backwardness of Brazil as its rigid internal structure which was created by colonialism. And that long after the demise of colonial rule, the traditional sector of the Brazilian economy was not really abolished, and was still disconnected from the real sector of the economy, thereby causing continued economic inefficiency and stagnation.[18]

Like Furtado, Sunkel also abandoned his earlier economistic analysis of the underdevelopment of Chile in particular, and Latin America generally. In agreement with Furtado, he was convinced that in order to gain a fuller idea of the economic problems that Latin America was faced with, its internal structure must be thoroughly understood. Thus, while he was optimistic that, for proper capitalist development as obtains in the United States of America and Europe to take place in Latin America, political alliances must be struck with the domestic bourgeoisie; and as a class, it must be organized in such a way to encourage active participation in production as against services, for the overall interest of the country.[19]

To both Furtado and Sunkel, the way out of the internal problems of Latin American countries rides with the radical reform of the internal structures that still have legacies of European colonialism. While it is true that the internal structures of the Latin American countries need a drastic change particularly the

the Brazilian economy had experienced some measure of growth and development, dependence is still deep-seated. And that since a lot of the critical production inputs like technology and capital were still supplied by the TNCs, the best that Brazil got was dependent development. In resting his case, Evans submitted that the era of classic dependence, which Africa is currently undergoing, was transcended in Brazil over 50 years, recommending that for the country (Brasil) to overcome its dependent development status, it needed a more powerful state which should unite with local capital to put the TNCs at bay.[28]

One of the major controversy surrounding Evans' thesis on 'dependent development' is that he portrays Brazil as a model for Africa whereas both of them are still faced with serious crisis of underdevelopment. This and other related issues caused Theotonio Dos Santos to join issues with him. In throwing more light on the pitfalls of Evans' thesis, Santos identified three forms of dependence which Brazil had undergone and is still undergoing: (i) colonial dependence; (ii) financial-industrial dependence; and (iii) technological industrial dependence.

According to Santos, while colonial dependence was characterized by trade monopolies complemented by the exploitation of land, mines and labour; financial-industrial dependence which, occurred in the latter part of the 19th century, was characterized by large concentration of international finance capitals in the centre with their investments directed into the production of materials and agricultural products in Brazil. The third, techno-industrial dependence, is characterized by the transnational corporations' establishment of branch industries to further reintegrate Brazil's economy into the global capitalist system. In his submission, Santos contented that, the lack of domestic capital, and the inability to integrate foreign capital with the local economy were not really the causes of the country's economic and industrial backwardness; instead, it was the imperialism of the TNCs. He recommended, therefore, a radical transformation of the internal structure.[29] Though Santo shed light on the shortcomings of Evans' thesis on dependent development, he ended up generating more problems. For instance, if one of the major problems for Brazil's backwardness was not lack of capital, then it would be difficult to imagine how the TNCs were to shape its direction of growth to serve their profit motive in the country. In transforming the internal structure of Brazil and other countries in the sub-region, Santos was also silent on the modalities to be adopted.

What came out very clear from the theoretical analysis of the above studies is that, in order to avoid sweeping generalizations on the role of the TNCs in the underdevelopment of the Third World countries in the future, country-specific studies such as the one undertaken here, is imperative. For, it is with such study that we can deepen our knowledge of the inter-connectedness between transnational capital, the state and local capital, and how together, their mode of capitalist exploitation underdevelops the host country. Suffice it to say that while it is true that the underdevelopment of the periphery has causal relations with the external, the dynamics of the internal structure particularly the character of struggle between and within the groups that make up the state plays a key role in the pillage of the host-countries. There is no doubt, that the TNCs have, through their operations, underdeveloped their host-countries in the South. However, the depth of the pillage can be more severe in some countries than in others. There is every tendency that the level of wreckage will be more in host countries where the state colludes with foreign capital than in others where it opposes. Which is why Shell can afford to operate in the Niger Delta with utmost recklessness since the Nigerian state willingly collaborates with it. Therefore, rather than holding the TNCs almost solely responsible for Africa's underdevelopment Rodney *et al* would want us to believe, a lot more is really wrong with the African state and its allies.[30] Let us now extend the frontiers of the analysis of these basic theoretical issues to the political economy of Nigeria in order to deepen our understanding of how Shell and the state underdeveloped the Niger Delta.

2.3. Basic Theoretical Issues in the Political Economy of Nigeria

Three basic theoretical issues have always dominated discussions on the political economy of Nigeria. They are the transnationals, represented by the state, and the domestic bourgeoisies. Together, they (three) have been indicted as chief culprits responsible for the underdevelopment of the country. However, in most of the received literature on the field, there is hardly concrete analysis of their exact impact as the country's economy, leading to sweeping generalizations; a problem that also pervades the majority of similar writings on the TNCs as earlier shown. In this section therefore, our main object is to critically examine and analyse these unsettled theoretical issues in order to provide a very rigorous basis for gaining a fuller knowledge of the role of Shell and the state in the underdevelopment of the economy and environment of the oil-bearing host areas in the Niger Delta. Let us begin the discussion with the state since it is central to the initiative and drive, wholly dependent on government and foreign capital, and incapable of fulfilling its historical mission of raising the productive powers necessary for the desired development of the people.[16]

Drawing on the experiences of Chile, another major host country to the TNCs in the sub-region, Osvaldo Sunkel demonstrated in his study how the country's economy was pillaged and brutally undermined by foreign capital.

Prebisch, Furtado, Cardoso and Sunkel identified one common problem with the economy of Latin America: stagnating and increasing dependence on the United States' TNCs. While these scholars might be right in claiming that the transnationals were the root cause of the stagnation of the economy of Latin America, they did not provide a rigorous basis for understanding its underdevelopment. As colonies of European imperial powers, the nature of the post-colonial economy of Latin America has its origin in colonial capitalism. So, too, is the nature of the state in post-colonial Latin America. Yet, Prebisch *et al* failed to explain the dialectics of history, and the politics of the state in Latin America in the period since political independence was attained. Their inability to capture these crucial issues, resulted in the recommendation of wrong remedies. Which is why the adoption of import substitution industrialization (ISI) as a potent antidote to reversing the sub-region's underdevelopment, ended deepening its integration into the world capitalist system. In fact, the experiences of Latin America with the ISI under the aegis of the TNCs showed that, rather enhance,it deepened its incapacity to produce locally, hitherto imported goods. The ultimate result is the import reproduction instead of import substitution of previously imported goods. Suffice it to note that, it is enough to describe Latin America's dependence on the US TNCs, and its overall underdevelopment as Probisch *et al* did.

Therefore, the early origins of the dependency school proper can be traced to the contradictions inherent in the propositions advanced by the structuralists in their studies. Contradictions that eventually gave rise to a group of scholars whose main concern was to address the origins and dialectics of the underdevelopment and dependence of Latin America and other peripheral social formations. But before going into a detailed discussion of the basic propositions of this group of scholars, it is important to note that, since Latin America had about the greatest concentration of United States and many Western European based transnationals among other developed countries in the 1950s and 1960s, most of the early studies on the TNCs in the Third World were drawn largely from the experiences of the sub-region. This explains in part why most of the analyses in this study also from studies on Latin America. It is not to assume, however, that studies have not been done on the operations of the TNCs in Africa; rather, they are just beginning to emerge. Now back to the discussion proper.

2.2.2.2. Underdevelopment and Dependency Theorists

To begin with, there are as many theorists on the underdevelopment and dependence of the Third World as there are varied propositions. Of importance to us in this study, however, are the common crucial issues in the received literature that underline the political economy of underdevelopment and dependence of social formations in the Third World. Issues like the state and its mode of surplus extraction, foreign capital, local private capital, and domestic bourgeoisie.

One of the most influential structuralist who to first radicalize the ECLA analysis of why Latin America was underdeveloped, was Celso Furtado. As Director of Planning of the poorest regions of Brazil, he personally experienced the development of events that led to the military take-over of the Brazilian government in 1964. To him, lack of capital was not so much the cause of the backwardness of Brazil as its rigid internal structure which was created by colonialism. And that long after the demise of colonial rule, the traditional sector of the Brazilian economy was not really abolished, and was still disconnected from the real sector of the economy, thereby causing continued economic inefficiency and stagnation.[18]

Like Furtado, Sunkel also abandoned his earlier economistic analysis of the underdevelopment of Chile in particular, and Latin America generally. In agreement with Furtado, he was convinced that in order to gain a fuller idea of the economic problems that Latin America was faced with, its internal structure must be thoroughly understood. Thus, while he was optimistic that, for proper capitalist development as obtains in the United States of America and Europe to take place in Latin America, political alliances must be struck with the domestic bourgeoisie; and as a class, it must be organized in such a way to encourage active participation in production as against services, for the overall interest of the country.[19]

To both Furtado and Sunkel, the way out of the internal problems of Latin American countries rides with the radical reform of the internal structures that still have legacies of European colonialism. While it is true that the internal structures of the Latin American countries need a drastic change particularly the colonial policies and structures of exploitation that were inherited, from the imperial powers, it is not too clear how the change will be brought about. For instance, what type of internal reform Latin American governments were expected to introduce, was not clearly spelt out.. There is no doubt that the state needs restructuring as they have suggested. But the question is how it would be done?. So, too, is the bourgeois class, which, as a colonial creation, is still inept politically and content with its role as a go-between for foreign capital in the local economy. For one thing, any reform that will threaten the existing political arrangement and mode of surplus extraction in the economy which favours the managers of the state and local allies of foreign capital will be resisted. As Lynn Mytelka demonstrated in her study on the crisis of regional integration in Latin America, the Andean Pact Foreign Investment Code of December 1970 failed partly because of the collusion of the state and local private capital with the TNCs to resist the overthrow of the status quo.[20] In essence, the state and its local allies are themselves, not really ready to change the policies and structures of capitalist exploitation.

The worsening overall decay in the internal economy of Latin American countries was to become of research interest to scholars like Paul Baran. According to Baran, the backwardness of the sub-region is directly tied to the continuous exportation of the economic surplus by the TNCs to the United States of America and Western Europe through profit repatriation and over-invoicing. And that even where surplus was not drained to the North, it was absorbed in unproductive ventures in the host countries.[21]

Building on Baran's analysis, Gander Frank argued that, underdevelopment is not an original state, but the concrete reflections of the past and present relations between the West and the Third World in which surplus has constantly been transferred from the latter to the former. To Frank, although the North may have been undeveloped before, it has never been underdeveloped in the context of the Third World countries of today. In his metropole-satellite framework, he explained how surplus was transferred in the world economy, an economy in which the North is the world's metropole whereas the South occupies satellite position. He demonstrated in his study that, the metropole-satellite relationship also existed between cities, regions, local centres and rural areas, as well as between landlords, merchants, peasants and labourers in the satellite. In this process, surplus is continually extracted outwards and upwards from one satellite to another metropole until it finally gets to the metropole of the capitalist world economy in the North.[22] In conclusion, Frank submitted that, once a country or a people is converted into the satellite of an external capitalist metropole, the exploitative metropolitan-satellite structure quickly comes to organize and dominate the domestic economic, political and social life of the people.[23]

Samir Amin also alluded to the center periphery construct of Frank. Together with Baran and Frank, Amin did not see the prospects of proper capitalist development as it obtains in the North, taking place in the South. This is all the more so as capital generated in the South is quickly siphoned to the North, which has hindered the capacity of the former to accumulate locally. And with its deepening incapacity to raise capital, the South can hardly undergo any steady growth in its economy much more of development as the prospect for capital development is already blocked.[24] Both Pierre Jalee and Walter Rodney shared Amin's position.

While for Pierre Jalee, the prospect for Third World countries generally to industrialize under the dependent relationship that binds them with the North is limited[25], Walter Rodney, on his part, traced the origin of Africa's underdevelopment in particular, to the 16th century when the European trade in human beings first helped depopulate the continent, and later, colonialism laid the basis for its present political and economic stagnation, and social decay.[26] And to Fitt, Faire and Vigier, what actually passes for foreign aid is no less than an imperialist mechanism for draining the resources of the South which reinforced its dependence on the North.[27]

Notwithstanding the useful insights gained into the political economy of the underdevelopment of the Third World countries, there are still confusion and sweeping generalization in most of the theoretical issues raised in the above studies. Firstly, is the contention that, proper capitalist development of the countries in the South is blocked. Which means that the path the West took to development is still taken as an ideal. And if that is the case, it all means that the development of the North will continue to shape that of the South. This is circulatory. Secondly, that the root cause of the underdevelopment of the South is still colonialism capitalism even long after most of the countries had gained political independence. I think a lot more is wrong with the post-colonial state and its path to development than blaming coloial capitalism solely for the rot in the economies of the South. Let us illustrate more on this with the study by Peter Evans in which most of these contradictions were reinforced.

Concerned with the relationship between the Brazilian state and the TNCs, Peter Evans examined the consequences of their activities on the country's economy. He demonstrated that, though process of surplus extraction in the country including its petroleum industry.

Foremost and leading theorist on the political economy of Nigeria. Claude Ake, has, in his various works, argued that, the Nigerian state is not a public force; nor is it objective in the sense that it uses its monopoly of coercion to police and guide the society rather impartially; nor does it manage the public institutions and resources to the interest of the public. On the contrary, it is a specific modality of class domination, and one in which class domination is mediated by commodity exchange so that the system of institutional mechanisms of domination is differentiated and dissociated from the ruling class and even the society, and appears objective standing alongside with society. That is, the state mode of domination is the way in which class domination is autonomized. In terms of its features, Ake argued that, the Nigerian state has little or no independence from the social forces of the society particularly the hegemonic class;and that as an imperialist project created out of the United African Company (UAC), it is a mere tool of foreign corporate capital, charged with the coercion of Nigerians into commodity relations, changing their pattern of production, and preventing the emergence of a competitive domestic bourgeoisie.[31] Nigeria's underdevelopment, Ake submitted, is to be found in these features whose implications for the country's economy are that:

> political and economic competition becomes normless, or at any rate, conducted in clear preference of efficiency to legitimacy norms. Contending groups struggle on grimly, polarizing their differences and convinced that ability to protect their interests and to obtain justice is extensive with their power. This creates politics of anxiety and under such circumstance, development cannot occur.[32]

Claude Ake might be right in claiming that the Nigerian state, as an imperialist creation, lacks autonomy, and therefore, cannot exercise an independent judgment over the development of the country. Doing so, meant assigning one main function to the state, is that of facilitating the capitalist accumulation of the country by both itself and foreign capital. If it is agreed that the Nigerian state is central to the process of accumulation process in the country's economy, the major question that arises then, is whether it does not need some autonomy to be able to perform this function effectively. No doubt, it does, and indeed, enjoys some autonomy which has assisted it to organize the different fractions of the ruling bloc into a conscious category, cater for the interests of the domestic bourgeois class, prosecute its mode of extracting economic surplus from the economy including the petroleum industry. Evidence of its relative autonomy abounds and in particular, stems from the political ineptitude of the domestic bourgeoisie. Since the domestic bourgeoisie is incapable of running the apparatus of government, but only concerned with mediating capitalist accumulation, the state cannot but exercise its own initiate on how to run the polity. That is not all as it needs autonomy to be able to effectively disorganize the dominated class, while acting as a representative of the ruling class whose interest it guarantees, and strengthening its own legitimacy and conditions that help reproduce the structures of the global capitalist system of which it is an intricate part. No other sector of the Nigerian economy have these issues been so clearly demonstrated than in the country's petroleum industry. More light will be shed on this in the course of the discussion.

In claiming that development cannot occur under the conditions of what he, Ake regarded as 'the politics of anxiety', it is still not too clear the type of development he really meant: capitalist or socialist?. With such suspense in his analysis, one is left wondering particularly in coming to grips with the nature of Niger Delta's underdevelopment. Finally, reducing the Nigerian state merely to an exploitative instrument of foreign capital used to meet its exclusive needs does not really provide a rigorous theoretical basis for accepting Ake's thesis on the autonomy of the Nigerian state. As noted earlier, the state is first and foremost, capitalist in all intents and purposes, and therefore, plays the crucial role of balancing all fractions of local capital in order to ensure that state-transnational capitalism in the economy thrives. This is particularly evident in the nature of the alliance between the forces of state capitalism and Shell in the country's oil industry.

Like Ake, Terisa Turner has,in her studies on the Nigerian political economy, indicted the transnational corporations and the state for the rot in the country. Characterizing the country's economy as basically export import-oriented, Turner identified the local business class as essentially on less than fronts for the transnationals, while the state itself, is a principal market being the major consumer of imported goods. As a main consumer in its own right, the grim struggle for the political control over the state and its apparatus has become an important source of making profit. As a result, the struggle for position in the state, or at least, access to those who exercise influence over political decisions, has become the ultimate essence of Nigerian politics. Turner also established a 'triangular relationship' between the transnational corporations, the state and the indigenous compradoral class which she preferred to call the 'collaborationist'; and that the international competition for accumulation among the TNCs, is being reproduced at the Nigerian local level through this triangular relationship. This proliferates especially as foreign capital intensified its bid to win lucrative contracts thereby leaving the country's economy in greater ruins. In fact, it is in this multiplicity of the 'triangular relationship' that the instability of the Nigerian state, which, in turn, structures the country's backwardness, is to be found. Turner further contended that, the prevalence of imperfect forms of competition between and among foreign and local capitals coupled with the absence of institutionalized forms of making profits, have transformed the Nigerian business class into being more concerned with having contacts in government instead of engaging in actual entrepreneurial ventures and real production. As a consequence, local entrepreneurs do not really organize labour, capital, raw materials and energy to produce for the market since there are easier, more profitable and less risky commercial alternatives for accumulation in the country. And finally, she argued that, the implication for Nigeria of the dominance of compradoral interest has been the prevalence of corruption and instability, and the inability of the state to organize the transfer of oil technology at a time when some members of the Organization of Petroleum Exporting Countries (OPEC) like Algeria has succeeded in doing so by finding, wining and refining crude oil almost independent of the oil TNCs.[33]

Turner might be right in her claim that, the Nigerian state is mainly a trading one is concerned primarily with the export and import of goods and services. However, the real issue at stake is not so much with the state being a merchant as its impact on the country's industrialization, particularly in the manufacturing sector and the capital good sector. Such obvious lacuna in her analysis fails to bring into bolder relief, how state commercial capitalism has either enhanced or impaired the reproduction of the world process of capitalist accumulation in the country. Corruption, is no doubt, a prevalent feature of the political economy of Nigeria. Again, she failed to elaborate on the logic behind it in the broader context of the accumulation process in the economy, and the reproduction of the capitalist mode of production not least that in the country's petroleum industry, a sector she maintained that the technology is still foreign.

Conceptualizing the domestic bourgeoisie and Nigerian politics purely at the level of exchange, as Turner did, neglected the various incentive measures which the state has articulated to encourage the transnationals and the indigenous entrepreneurs to invest in production as against mere exchange. Even in limiting Nigerian politics to the struggle for commercial gains, she failed to bring out very clearly, the nature of political competition in the country. For, beyond exchange, the grim struggle among and between the feudalists, the military and local capital in the polity especially over the control of the apparatus of the state has shaped the bloody character of Nigerian politics. So, too, has ethnicity structured Nigerian politics especially the three major ethnic groups of Hausa/Fulani, Yoruba, and Igbo which have structured the configuration of political power in the country. For instance, as the three ethnic groups compete fiercely with each other over who should control the government at the centre where the final political power lies, the minority groups within these major ethnic zone have also tried to redress their marginalization by way of either withdrawing their support or breaking away from the polity. A case in point is the *Ogoni Bill of Right* in which the people of Ogoniland wanted a state for themselves since they still live in abject poverty in spite of being a major oil producing area since 1958. These and other neglected issues in Turner's analysis are too central to a deeper understanding of the character of Nigerian politics to be neglected. In fact, it is these issues that largely define the politics of environmental degradation in the Niger Delta.

Like the state, the national domestic bourgeoisie has been variously condemned for being partly responsible for Nigeria's underdevelopment. Branded as unproductive and acting as a 'transmission belt' for transnational capitals, Segun Osoba argued that, the domestic bourgeoisie is tied to Euro-American finance capitals in an agent/principal relationship in both the colonial and post-colonial periods. Osoba's main contention is that, the domestic bourgeoisie is not unaware of its subordinate status; instead, he advanced four main reasons for its inability as a class, to reverse its junior partner relationship with foreign capital.

First, the lop-sidedness of the international division of labour reinforced the refusal of the transnational corporations to transfer technology to Nigeria, or split ownership and control of the economy with Nigerians. All this has acted to keep the national bourgeoisie in its place as a subordinate class. Second, the Nigerian national bourgeoisie found its role as a 'commission agent' and membership of the board of directors of the TNCs' subsidiaries operating in Nigeria, lucrative since it helps its class members to further their selfish economic interests citing the cement 'armada' and the 'scania bus scandal' as concrete examples. Third, the cohesion of the national bourgeoisie itself, is weak as its unity is undermined by fractional differences. This has placed it at a disadvantage position in its negotiation with the transnational corporations. Fourth and final, is the contradiction between the national bourgeoisie and the Nigerian masses; contradiction being engendered by the unwillingness on the part of the national bourgeoisie to end its dependence on the TNCs, create an egalitarian society in which the masses will be favoured despite its defensive radical positions on the one hand. And on the other, to admit openly to the masses that its true commitment is to assist its foreign masters, that is, the TNCs, to facilitate the exploitation of the country. In resting his case, Osoba insisted that the root cause of Nigeria's underdevelopment is located in the inability of the national bourgeoisie to retain and use an adequate proportion of the nation's wealth that it has appropriated for the upliftment of the material well-being of the masses. This failure on its apart,

> has robbed succeeding regimes in the country of legitimacy in the eyes of the generality of Nigerians, pointing to the high perishability of governments and the unstable and volatile nature of the state as a function of the resultant legitimacy/ authority gap. The only way out of the crisis according to Osoba, is to create a superior people's revolutionary culture and organization to overthrow the national bourgeoisie.[34]

The impression that Osoba has created in his study is that, the domestic bourgeoisie as a class, contributed to the underdevelopment of the Nigerian economy. That is well taken. But its role in the pillage of the country was not in a magnitude that he, Osoba, wanted us to believe. On this score, Bjorn Beckman disagreed with Osoba's analysis of the domestic bourgeoisie. Osoba's location of the conflicts between the domestic bourgeoisie and the Nigerian people merely as a function of the former's role as an agent of foreign capital and not because it is a bourgeois class, was rejected by Beckman. This prompted Beckman to ask Osoba that, if the national bourgeoisie were to be more nationalistic and ceased to be agents, would a stable, efficient and legitimate government be established, and squalor, congestion and mass poverty reduce in the country? To him, the reality of the situation is that, both the bourgeoisie and the people are on the same side of the contradictions arising from transnational capitalism.

On the bourgeoisie's siphoning of the nation's a wealth abroad and its negative consequences for the Nigerian economy, Beckman raised serious doubts as to whether if the siphoning were to stop, it will necessarily be used for the upliftment of the toiling masses of Nigeria. Pointing to the huge economic resources made available to the Nigerian state after the OPEC oil price increases of 1973, he argued that, rather than eliminate poverty, social inequality, injustice and the material well-being of the majority of Nigerians worsened. Thus, although the outflow of profits from the Nigerian economy through the bourgeoisie is indicative of the inefficiency of the state, they do not provide enough evidence to justify the discussion of national development and class struggle in the way Osoba did. If the state and the bourgeoisie are held responsible for the country's backwardness mainly because they are weak and not performing, Beckman's contention is that, where will he, Osoba, locate his thesis when they start to perform; or should it be assumed that they will never perform?[35]

Julius Ihonvbere's study of the role of the national bourgeoisie in the underdevelopment of Nigeria is even more critical than Osoba's. He asks:

> Which bourgeoisie? You mean those fools? Just look at them... Is that how their counterparts in the countries they are copying behave? They lack discipline. They are corrupt. They are agents to the multinationals. They cannot even practise politics in a peaceful and responsible way. Unless they manipulate the poverty and differences of the people, they feel unsafe. What are they really good at? Look at the conditions of our people and country since flag independence in 1960, you will see how the Nigerian bourgeoisie subverts its own future. I cannot, therefore, call them bourgeoisie.[36]

Extending Beckman's criticisms on Osoba to Ihonvbere, it is not really clear what he, Ihonvbere meant by characterizing the national bourgeoisie as not good at anything. For, the main issue at stake is who a bourgeoisie is in the first place, and how the Nigerian bourgeoisie is really different in terms of its location in the process of surplus extraction and politics from its counterparts in the developed capitalist state of the North? For, the bourgeoisie, as a class, is primarily concerned with the reproduction of the capitalist mode of production in order to strengthen its own accumulative base whether in Nigeria or in the developed economies. To expect it to be nationalistic in spite of its political ineptitude as Ihonvbere and Osoba would want us to believe, is placing undue emphasis on the ordinary. Ihonvbere's logic of how the activities of the bourgeoisie 'subvert its own future' is rather not very clear. As part of the state structure and its state mode of capitalism, all that it (domestic bourgeoisie) is concerned with, is survival. And it cannot achieve this by 'killing' the existing process and nature of capitalist accumulation in the economy. Besides, no insight is really gained into how the non-performance of the bourgeosie as a class, has hindered its ability to mediate the reproduction of the global capitalist accumulation. Perhaps, if the studies by Osoba and Ihonvbere were sector-specific, most of the generalizations and the counter-factualizations that dominated Beckman's criticisms would have been avoided. For instance, the politics of the degradation of the environment in the oil areas of the Niger Delta and the resultant protest by the people who live there as shall be revealed in the course of the discussion, will help bring out very clearly, how the domestic bourgeoisie, though at the root of the unrest in the oil-bearing communities, is only doing so for its economic gains, and not that it really wanted to halt environmental pollution.

CODESRIA's Group Study on Nigeria's development demonstrated that, proper capitalist development similar to what obtains in the advanced capitalist states of the North, is unattainable in the country the major reason being that it is a dependent on the TNCs and a operates a mono-product economy. Okwudiba Nnoli, who led the group project, argued that, a lot of what passes for technological development in the country was no less than mere collection of Western 'artefacts' in the name of technology, while the country's economy remains externally-oriented to serve the demands of foreign capital. To him, the technology being imported into the country is inappropriate; rural areas continued to suffer from neglect and decay; and the economy is still dominated by western finance capitals without a corresponding domination of western economics by Nigeria's indigenous corporate capitals. Bade Onimode, another contributor to the CODESRIA sponsored study, argued that, industrialization under the aegis of the TNCs is not possible in Nigeria as their profit motives are incompatible with the development objectives of the state. That colonial capitalism took active measures to prevent manufacturing in Nigeria in order to guarantee its sources of raw materials and market outlets for the products of the British transnational capitals. The positions of other contributors to the study on the backwardness of Nigeria were not different from those of Nnoli and Onimode and need be repeated. The consensus of opinion among them was that, 'capitalist development as it obtains in the West, is not possible in Nigeria', suggesting in the alternative, a socialist path to development.[37]

In claiming that 'proper capitalist development' is not possible in Nigeria, the CODESRIA Group takes the experiences of the West as given, irrespective of the fact that it was never underdeveloped like Nigeria. In other words, they would have preferred Nigeria follow the path of development undertaken by the West. That, is itself, a contradiction. Even the socialist path to development advocated by the contributors did not provide how it will be achieved. Should it start with the state? This is similar to the suggestion by Sunkel, and Furtado. Furthermore, while it is true that TNCs do underdevelop their host countries in the Third World, it is not very clear how they (TNCs) through their operations, exactly undermined Nigeria's capacity to industrialize in the CODESRIA study. For, the aim of the capitalist development the TNCs have promoted in the Nigerian economy is to guarantee and expand markets, and maximize profits. No doubt, some level of industrialization can take place, but the problem is the type and interest served. This is the situation in Nigeria. Rather than genevaline, specific studies on the activities of transnational corporations such as the one undertaken here, will provide a rigorous basis for bringing into the fore, a clearer picture of the extent and nature of the underdevelopment of Nigeria. No doubt, technology is about the most effective tool used by the TNCs to expand their imperialism in the global capitalist economy. But this is not without the consent of the state whose mode of capitalism is in tandem with the transnationals' interests. So, if the technology being transferred to Nigeria is no less than artefacts as Nnoli argued, it is all part of the collusion between the state and the transnationals. Furthermore, any technology is designed to serve a particular purpose. Nnoli's analysis did not shed light on this, especially the use into which the artefacts were put, and its negative consequences for capitalist accumulation in the economy, particularly its deepening industrial backwardness.

Onimode's claims that colonial capitalism prevented real manufacturing from taking root in the country is well taken. Yet, his contribution to the study did not provide much basis for reaching such conclusion. So, too, was his submission that industrialization in Nigeria was blocked. The real issue is not so much that TNCs are promoting industrialization as evidenced in the case of Brazil's experience with dependent industrialization; rather, it is the type and its implication for the economy of the host state. This was obviously not part of Onimode's thesis on Nigeria's 'blocked industrialization'.

One of the main issues raised in Beckman's criticism of Osoba analysis, is that so much money in the form of rent accrues to the state; yet, it is unable to use the funds to reverse the country's industrial and economic underdevelopment and uplift the material well-being of the majority of Nigerians. The failure on the part of the Nigerian state to use the wealth at its disposal to bring about the country's economic development interested Tom Forrest. To Forrest, it is because the state is rentier, explaining its preference for the collection of royalties and rents accruing from the oil companies and proceeds from oil export without judiciously investing them in production.[38]

Forrest might be quite correct in claiming that the Nigerian state is a collector of rent especially in the light of its role in the operations of the country's oil and gas industry; an industry where the state still relies exclusively on the oil transnationals to produce crude oil. However, Forrest's thesis on the rentier state fails to explain the uses into which the rent is being put. For, all capitalist states whether in the North or South, collect rents and taxes in one form or the other, which are then invested to reproduce the global capitalist system. Rent therefore, serve as a means to an end. Branding a state rentier as Forrest did, simply because it collects rent, without giving an insight into what the rent is being used for, especially as to whether it helps reproduce the process of capitalist accumulation at both the local and external levels or not, is rather misplaced.

In his comparative case study of Nigeria, Pakistan and the Philippines, Akeredolu-Ale gave useful insight into the predatory and exploitative orientation and activities of the TNCs, their inherent tendency to resist and hamper local industrialization and perpetuate mercantile capitalism, and their deliberate efforts to dwarf the rise of indigenous entrepreneurship in these countries. He contended that, the presence and activities of the TNCs have hindered, rather than helped the development of a local business class in Nigeria. So, too, was the situation in Pakistan and Philippines. In the particular case of Nigeria, he suggested that the only way out is for the government to establish industrial and commercial banks to cater primarily for the interest of the local businessmen.[39]

In holding foreign capital responsible for the inability of the local business class to really emerge in Nigeria, Akeredolu-Ale has assumed that foreign capital is synonymous with *underdevelopment*, and that local capital means real development in Nigeria. Obviously, this cannot be so. For, though foreign capital and the domestic bourgeoisie are unequal partners, both of them help reproduce the global capitalist system by exploiting the people. Suffice it to note that, the empowerment of the local business class as Akeredolu-Ale has suggested, might not necessarily bring about the type of development that he advocated for Nigeria. It is all the more so since he did not metion what should be done to the state, especially as it is central to the process of industrial accumulation.

The indigenization decrees of 1972 and 1977 promulgated by the Nigerian state were aimed at wresting control of the commanding heights of the country's economy from foreign capital and it in the hands of local private capital. But the promulgation and implementation of the Decree have sparked off a lot of argument on the political economy of underdevelopment of Nigeria. First, is a group of scholars, who conceptualizes the Nigerian state as purely an imperialist creation, arguing therefore, that, its indigenization decrees are nothing but measures to reinforce the lopsided partnership between itself, foreign capital, and its local allies. Notable among this group, is Segun Osoba, whose contention is that, the indigenization decrees are bound to be ineffective because the TNCs are not ready to split ownership and control with the domestic bourgeoisie. To Claude Ake, another great critic of the decree, it made more Nigerians owners of the indigenized companies but absolutely without control. In the area of technology, Ake maintained that the foreign firms still calls the shot. Other critics of indigenization like Bade Onimode contended that, what ought to have been a collective national response to the imperialist domination of the country's economy turned out to be a wholesome failure.[40] While these criticisms are well taken, there were no detailed specific case studies to really validate most of these general claims. Perhaps, this was to prompt Ankie Hoogvelt to undertake a study of Kano.

Drawing on her experience of the impact of the indigenization decrees on the Kano economy, Ankie Hoogvelt argued that, the country's oil money has been used to finance the growth of firms which have remained under foreign control. To her, many of the transnational corporations' subsidiaries that were indigenized, merely hand-picked their shareholders, who were later used as 'front men' to lobby government officials for land acquisition, contracts, expatriate quota among others. The ultimate result was that foreign capital, still retained control over Kano's economy.[41]

That Kano's economy is further reinserted into the global capitalist economy under indigenization as Hoogvelt's analysis shows, is not open to dispute. But the findings from the Kano case cannot hold for other sectors of the country's economy. In the oil and gas sector, particularly in its upstream, the oil TNCs whether during indigenization and after, are still wholly owned by their parent companies in Europe and the USA. That explains in part why the control of the oil and gas sector still resides with the oil TNCs. Little wonder why Olusegun Obasanjo, then head of the Federal Military Government and whose regime put the 1977 decree into force, owned up that the indigenization decrees were a disaster. His own words:

Although the economy may be said to be largely indigenized, what has happened is that, Nigerians have taken over rights to share in profits while control has remained where it has always been, in foreign hands, which, if we must be realistic, cannot reasonably be expected to identify as intimately with national objectives as if the situation were otherwise. Clearly, control of the commanding heights of the economy which the decree had, as one of its main objectives, is nowhere in sight... and indigenization may have missed the substance for the shadow.[42]

The other group of scholars has, as its main contention, that the indigenization project of the Nigerian state helped the domestic bourgeoisie advance its economic position vis-a-vis foreign capital. Leading writers of this school, notably Paul Collins, has emphasized that, the indigenization of the Nigerian economy made it possible for the domestic bourgeoisie to harness state power to its own advantage by using government regulations to weaken the control of the country's economy by foreign capital. Based on his study of Kano and Lagos, Collins, in contrast to Hoogvelt, demonstrated that the local shareholders derived much benefits from indigenization, pointing to the fact that a large number of local entrepreneurs acquired shares in several manufacturing companies whose operations were closely connected with their own line of business and technology. He also showed that the intention of the local businessmen was to create linkages between the foreign firms and a local assembly plant that they planned to establish; and that in enterprises where indigenous businessmen held majority shares, they ceased to be subsidiaries of the TNCs, and were accordingly, charged for any input supplied to them from abroad. In his submission, Collins rejected the thesis on neo-colonial state which the first group of scholars developed on indigenization; and offered instead, that the state should be seen as a register of the balance of class forces with the domestic bourgeoisie holding the leading edge.[43]

Bjorn Beckman rejected the interpretations given by the two groups of scholars on the indigenization decrees as inadequate. On the employment of the 'neo-colonial model' in the analysis of the indigenization project, although Beckman agreed that there is evidence to support the claim that the economy was worse-off, the Nigerian state is not a mere creation of one faction or fraction of foreign capital or the other. Alluding to such position, Beckman insisted, would give the impression that the state only functions to meet the absolute interests of these capitals. In agreement with

Beckman, my contention is contend that such position has robbed the domestic business class of its strength and organization. Referring to Hoogvelt's claims that the unpatriotism on the part of the local businessmen was responsible for the failure of the indigenization decrees, the impression she created is that should it be patriotic, all will be well.[44]

With respect to the economic nationalism of the Nigerian domestic bourgeoisie under indigenization, Beckman picked hole in Paul Collins' argument that it is anti-imperialist. Rather, for its own class reasons, it merely wanted a bigger share of the TNCs profitable business operations in the country; and that both local and foreign capitals made gains from indigenization. Beckman insisted that all that the state did, was to establish, maintain, protect and expand the conditions of capitalist accumulation in the country, without which neither foreign nor Nigerian capital can prosper. In other words, it (state) played the double role of enhancing and consolidating the domestic bourgeoisie to enter into a close partnership with foreign capital, while at the same time, strengthening the credentials of foreign capital as national Nigerian companies.[45]

The points raised by Beckman are well received. However, his argument that the state was charged with the primary duty to ensure the prevalence of an enabling environment for capitalist development under indigenization, is doubtful. For, together, the domestic bourgeosie and the people are invariably on the same of exploitation by the TNCs. Although his (Beckman) logic of state capitalism deepens our understanding of the role of the state in the indigenization project, his analysis led us to believe that the state is class neutral even when it is nothing of the sort. Furthermore, his construct on state model of capitalism portrayed the state to be enjoying unlimited autonomy, which, it is not the case in reality particularly in the high technological areas like the oil and gas sector where the indigenization decrees were silent since the state is wholly dependent on the TNCs for oil exploration and production. The question that arises therefore, is how will he (Beckman) explain his seeming unlimited autonomy of the state when it cannot produce crude oil independent of the oil TNCs long after the indigenization decrees came into force before they were repealed in 1997 by the late General Abacha junta. Details on these issues are examined in the light of Shell in Nigeria in the course of the discussion.

2.4. Concluding Remark

No doubt, deep insights have, been gained into the political economy of the transnationals in the developing economies from the analysis of the above studies. As earlier pointed out, a lot of the conclusions were reached in most of the studies, came not from specific studies, but general surveys. It is in order at this point to examine and analyse studies on the country's oil and gas sector in order to further deepen our understanding of the unsettled theoretical issues underpinning Niger Delta's underdevelopment by Shell and the State. It is this task that Chapter Three sets out to address.

Notes and References

1. For details, see Louis Turner *Invisible Empires*, (New York: Harcourt, Brace, Jorarroviet) 1971; and his *Multinational Corporations and the Third World*, (London: Allen Lane), 1974; Stephen Robock and Kenneth Simmoods, *International Business and Multinational Enterprises* (Homemad: Irvin), 1973; Mira William, *The Emergence of Multinational Enterprises* (Cambridge: Havard University Press) 1969; Jack Behrman, *National Interests and the Multinational Enterprises* (Englewood Cliffs, N.J. Printice-Hall) 1970; Charles Kindleberger ed., *The International Corporation... A Symposium*, (Cambridge: MIT Press) 1970 and his *American Business Abroad* (New Haven: Yale University) 1969; Gardney Brown ed., *World Business: Promise and Problems* (New York: Free Press) 1970; Edward McCreasy, *The Americanization of Europe* (Garden City: N.Y. Doubleday) 1964; and Kari Leritt, *Silent Structure: The Multinational Corporations in Canada* (New York: St. Martin's Press), 1970.
2. For works on development theory, see Walt Rostow, *The Stages of Economic Growth* (New York: Cambridge) 1960 and his edited volume, *The Economics of Take-Off Into Sustained Growth*, (New York: St. Martin) 1963. For particular works done by other developmentalist scholars on the tncs, see Charles Kindleberger ed., *The International Corporation: A Symposium*, (Cambridge: MIT Press) 1970; Thomas Bierstaker, *Distortion or Development? Contending Perspectives on the Multinational Corporation* (Cambridge: MIT Press) 1978; The United Nations Report titled *Multinational Corporations in World Development* (New York: United Nations ST/EC/190) 1973; Richard Barnet and Ronald Muller, *Global Reach: The Power of the Multinational Corporations* (New York: Simon and Schuter) 1974. Rayond Vernon, *The Economic Consequences of Multinational Enterprise: An Anthology*, (Boston: Harvard) 1972 and his numerous

works; Peter Drucker, 'Multinationals and Developing Countries; Myths and Realities', *Foreign Affairs*, 53) 1974; John Diebold, 'Multinational Corporation: Why be Sacred of Them?' *Foreign Policy 12* 1973.

For details on dependency theory, see Charles Wilber ed., *The Political Economy of Development and Underdevelopment* (New York: Random House) 1973; Gander Frank, *Capitalism and Underdevelopment in Latin America: Historical Studies of Chile and Brazil*, (New York: Monthly Review Press 1967; Paul Baran, *The Political Economy Growth* (New York: Monthly Review Press) 1957; Celso Furtado, *Development and Underdevelopment* (Berkeley: California) 1967 and various works done by Samir Amin, Ciovanni Arighi, Theotonio Dos Santos, Colin Leys, Immanuel Wallerstein. See also, Magnus Blomstrom, *Development Theory in Transition: The Dependency Debate and Beyond Third World Responses* (London: Zed Press) 1984; Stephen Katz, *Marxism, Africa and Social Class: A Critique of Relevant Theories* (Montreal: McGill Press) 1980, (monograph); Satchi Ponnambalam, *Dependent Capitalism in Crisis: The Sri Lankan Economy 1948 - 1980* (London: Zed Press) 1981; Bjorn Beckman's 'Imperialism and Capitalist Transformation: Critique of a Kenya Debate' *Review of African Political Economy* (ROAPE, 20-22) 1981.

3. Cited in Joseph Schumpeter, *The Theory of Economic Development* (Cambridge: Harvard University Press) 1934. Schumpeter studied the problems of growth and development.
4. Roy Harrod, *Towards a Dynamic Economics* (London: McMillian) 1948; and Evsey Domar, *Essays in the Theory of Economic Growth* (Oxford: OUP), 1957.
5. *ibid*, Harod and Domar.
6. Walt Rostow, *The Stages of Economic Growth* (Cambridge: Cambridge University Press) 1960.
7. *ibid*.
8. David Ricardo, *The Principles of Political Economy and Taxation*, (London: Pelican) 1970.
9. See Arthur Lewis, *The Theory of Economic Growth*, (London: Allen and Union) 1954; and in particular, his 'Economic Development with Unlimited Supply of Labour', *The Manchester School of Economic and Social Studies*, XXII, 2, 1955; and 'Industrialization of the British West Indies', *Carribean Economic Review*, 2, 1, 1950.
10. See Peter Drucker, 'Multinationals and Developing Countries: Myths and Realities', *op. cit*.
11. See David Blake and Robert Walters, *The Politics of Global Economic Relations* (Englewood: Prentice Hall) 1976.

12. Barnet and Muller, *op. cit.*
13. *ibid*, pp.215 - 216.
14. Raymond Vernon, *op. cit.*
15. See Raul Prebisch, *The Economic Development of Latin America and Its Principal Problems*, (New York: United Nations) 1950.
16. Celso Furtado, *Economic Growth of Brazil*, (Berkeley: Berkeley University Press) 1963.
17. Osvaldo Sunkel, 'National Development Policy and External Dependency in Latin America', *Journal of Development Studies*, 1,1, 1969.
18. Furtado's comment in 1966 and cited in A. Hirschman, 'The Political Economy of Import-Substitution Industrialization in Latin America', *The Quarterly Journal of Economics*, 28, 1968.
19. See Osvaldo Sunkel, *op. cit.*; and in particular, his 'Transnationalization and Its National Consequences' in J. Villamill ed., *Transnational Capitalism and National Development: New Perspectives on Dependence* (Sussex: Harvest Press), 1, 1969.
20. See Lynn Mytelka, *Regional Development in a Global Economy: The Multinational Corporation, Technology and Andean Integration*, (New Haven: Yale University Press), 1979. See also Daniel Omoweh, 'Integration and Collective Self-Reliance in Developing Countries: A Case Study of the Andean Pact', Unpublished M.Sc. Thesis, 1985, International Relations Department, University of Ife.
21. Paul Baran, *The Political Economy of Growth*, (New York: Monthly Press Review), 1957.
22. Gander Frank, *Capitalism and Underdevelopment in Latin America*, (New York: Monthly Review Press), 1967.
23. *Ibid.*
24. Samir Amin, 'Accumulation and Development: A Theoretical Model', *Review of African Political Economy*, 1, 1972; and his *Dynamics of Global Crisis*, (New York: Monthly Review Press) 1982.
25. Pierre Jalee, *The Third World in World Economy* (New York: Monthly Review Press), 1969; and his *Imperialism in the Seventies*, (New York: Monthly Review Press), 1974.
26. Walter Rodney, *How Europe Underdeveloped Africa*, (Dar-A-Salaam: Tanzania Publishing House), 1972.
27. Yan Fitt, Alexander Faire and Jean-Pierre Vigier, *The World Economy in Crisis: US Imperialism at Bay*, (London: Zed Press) 1980.
28. Peter Evans, *Dependent Development: The Alliance of Multinationals, State and Local Capital in Brazil*, (Princeton: Princeton University Press) 1978.

29. Theotonio Dos Santos, *Imperialism and Dependence*, (Mexico: ERA) 1978.
30. Gabriel Palma, 'Dependency: A Formal Theory of Underdevelopment or Methodology for the Analysis of Concrete Situations of Underdevelopments', *World Development*, 6, 7 & 8, 1978.
31 Claude Ake, *Political Economy of Nigeria* (London: Longman 1985), pp.1-3. For details, see his *A Political Economy of Africa* (London: Zed Press 1978). For other works on state in Africa, see for instance, Eme Ekekwe *Class and State in Nigeria*, (London: Longman 1986), John Saul, 'The State in Post-Colonial Societies: Tanzania'. *Socialist Register*, (1974), Colin Leys, 'The Overdeveloped' Post-Colonial States: A Re-evaluation', *Review of African Political Economy* 5, (1976) and his *Underdevelopment in Kenya: The Politics of Neo-Colonialism*, (Berkeley: University of California Press, 1975); and Steven Langdon, 'The State Capitalism in Kenya', *ROAPE*, 8, 1977.
32. Claude Ake, *Political Economy of Nigeria*, *op. cit.*, p.10.
33. T. Turner, ' The Transfer of Oil Technology and the Nigerian State', *Development and Change*, 7, (1976); and her 'Multinational Corporations and the Instability of the Nigerian State', *ROAPE*, January - April, 1976. See also, Garvin Williams ed. *Nigeria: Economy and Society*, London: Rex Collings, 1976.
34. Segun Osoba, 'The Nigerian Power Elite; 1952-1965' in Peter Gutkind and Peter Waterman eds., *African Social Studies: A Radical Reader*, (New York and London: Monthly Review, 1977); his 'Decolonization in Nigeria as a Programmed Transition to Neo-colonial Dependence', Seminar paper, Department of History, University of Ife, 1977; 'Considerations and Some Conceptual and Ideological Aspects of Nigerian Underdevelopment in Historical Perspective' in *Proceedings of the Department of History Seminar Series 1979/80 Session*, University of Ife, 1980; and 'The Deepening Crisis of the Nigerian National Bourgeoisie', *ROAPE*, 13, 1978. Also, see, Julius Ihonvbere, 'The State and the Irrationality of the Bourgeoisie: How the Nigerian Bourgeoisie Subverts its own Future', conference paper presented at *AWO: The End of An Era*, organized by Obafemi Awolowo University Press, Ile-Ife, October, 4-8, 1987; and his *Rise and Fall of Nigeria's Second Republic 1979 - 1984*, co-authored with Toyin Folola (London: Zed Press, 1985).
35. See Bjorn Beckman, 'Imperialism and the National Bourgeoisie: Critique of a Kenyan Debate', *ROAPE*, 1981, *op. cit.*
36. Julius Ihonvbere, *op. cit.*, 1987, p.1.
37. See Okwudiba Nnoli ed., *Path to Nigeria Development*, (Dakar: CODESRIA, 1981). Members of the CODESRIA study group are: Okwudiba Nnoli, Bade Onimode, Inyang Eteng and U. Nwala.

For other works done by Bade Onomode, see his, *Multinational Corporations in Nigeria*, (Ibadan: Les Shyraden 1983) *Imperialism and Underdevelopment in Nigeria* (London: Zed Press, 1982) and *The Political Economy of Crisis in Africa* (London: Zed Press 1989).

38. See Tom Forrest, 'Recent Development in Nigerian Industrialization' in Martin Fransman, *Industry and Accumulation in Africa*, (London: Heineman, 1982.
39. See Akeredolu-Ale, E., *The Underdevelopment of Indigenous Entrepreneurship in Nigeria*, (Ibadan, Ibadan University Press) 1980.
40. Segun Osoba *op. cit.* (1978); Claude Ake *op. cit.* (1978); and Bade Onomode *op. cit.* 1982 and 1983.
41. Ankie Hoogvelt, 'Indigenization and Foreign Capital: Industrialization in Nigeria', *ROAPE*, 14, January - April, 1979.
42. Cited in Claude Ake *op. cit.* 1985, p.189.
43. For details, see Paul Colins ed., 'The Political Economy of Indigenization in Nigeria', *Special Issue of the Quarterly Journal of Administration* (Ife), 9 January 1975; his 'Nigeria's Indigenization Policy' in Proceedings of the Nigerian Economic Society symposium, 1974, (Nigerian Economic Society, University of Ibadan Press, Ibadan, 1975; and *Survey of Indigenized and Partially Indigenized Enterprises, Lagos and Kano, 1975* submitted to the University of Ife.
44. Beckman, *op. cit.*, *ROAPE*, 20-22, 1981, 23 1983.
45. *Ibid*, 20-22, 1981.

Chapter 3

Oil Companies and Capitalist Development in Nigeria

3.1. Introduction: State-Transnational Capitalism and Underdevelopment of Nigeria

Studies on the Nigeria's petroleum industry are recent partly because the country's oil and gas sector is itself, still evolving. Like studies on the political economy of Nigeria, the basic theoretical issues raised in the received literature on the country's petroleum industry center on development and underdevelopment. Indeed, it is no less than an extension of the development and dependency schools of thought on the activities of transnationals, in the country's oil industry. The basic contention of the development school is that, the oil TNCs have, through their operations in Nigeria, provided the vital missing production inputs, such as capital and technology with which the Nigerian state began the development of its oil industry; and that the rent accruing from oil production and export was used to embark on the economic development of the country. A major proponent of this school is Scott Pearson, whose main contention is that, as the foreign oil companies provide investment capital, manpower and equipment they helped develop the host country's oil sector thereby contributing directly to its overall development. In tracing the history of Nigeria's petroleum industry and the role of the oil companies from the colonial period to the late 1960s, Pearson argued that, though their vertically integrated structure notwithstanding they impacted, positively on the country's economy citing tremendous revenue accruing from oil export, taxes and rents paid by the operating oil companies as evidences. Other benefits, according to Pearson, included the inflow of skilled expatriate labour from Europe and USA into the country's oil industry, which helped build the capacity of the local oil engineers and other related personnels in both the up- and down-stream sectors of the industry. Apart from these benefits, Pearson also noted some of the basic problems of the industry. Like Terisa Turner, Pearson was of the view that, Nigerian politics was nothing but the grim struggle among the institutional groups of the state for the control of state political power where the key to wealth accumulation in the economy resides. To him, the Nigerian civil war was just an evidence of such grim struggle over the political control of the state and indeed, oil wealth. He rested his case with a recommendation that the state should increase its regulatory role of the oil industry.[1]

Pearson's fundamental position on the political economy of the oil transnationals in Nigeria is not significantly different from that of Peter Drucker's on the transnationals generally. The criticisms earlier levelled against him (Drucker) in Chapter Two, are too familiar to warrant a repeat here. However, Pearson went further to raise other critical theoretical issues which are characteristic of other structuralists like Barnet and Muller. While it might be true that the vertical nature of the foreign oil companies facilitated the expansion of the imperialism of monopoly capitalism across national frontiers, Pearson failed to demonstrate its specificities and implications for the economy of the country inclusive of the oil industry. His inability to empirically situate such crucial issue in the study not only created an analytical gap, but which left one wondering on wht informed his evaluation of the cost and benefit of the inflow of direct foreign investment into the country's oil sector. So, too, was it difficult to ascertain how he then concluded that, the oil companies contributed to the development of the Nigerian economy bearing in that what may pass as benefits is actually not, but only a means to further the capitalist exploitation of crude oil.

It is not in doubt that the new-found oil wealth in Nigeria introduced a new dimension into the country's character of politics, particularly the intensification of the fierce struggle to capture and privatize the state at all cost and by all means by the political class. However, reducing political competition in the country merely to the level of fierce struggle by the political class to exercise absolute political control over the oil wealth as Pearson did, does not provide a rigorous basis for its understanding. For, ethnicity and minority agitations, and the military-civil society relations are too crucial an issue to be neglected in such discourse if a deeper idea of the nature of political competition in the country and its implications for democracy and development are to be gained. Therefore, Pearson's analysis of the causes of the Nigerian civil war, 1967 - 1970, purely at the level of political competition over the control of the country's oil wealth has placed much emphasis on the symptom of, instead of the root causes of the crisis. Finally, it is not too clear what Pearson really meant by the 'increased regulatory role of the state'. For, both during the period covered in his study and now, what the state still basically does in the oil and gas industry, is the collection of rent from the operating foreign companies. If that is what he meant, which is most likely to be, then he, Pearson, has endorsed a weaker role for the Nigerian state at a time when it is expected to wrest control of the country's oil industry from the oil transnationals.

In his study, Ludwig Schalz was particularly interested in how much revenue the host state ought to have earned if the foreign oil companies were to operate selflessly. AS to be expected, he accused Shell-BP of deliberately delaying the production of crude oil in Nigeria after the end of World War II in 1945 because of the company's huge oil deposits in the Persian Gulf.[2]

Schalz might be right in claiming that Shell-BP delayed oil production in Nigeria since it was still producing its large reserve in the Persian Gulf. The question that arises however, is whether Shell could have done otherwise particularly since its operation as obtains in any capitalist enterprise, is largely governed by profit maximization. For one thing, oil companies are known to develop oil reserves with full recourse to the trend of the global oil market and not on charity or political sentiments. For instance, the Oso Condensate currently being exploited by Mobil in Akwa Ibom State of Nigeria, was first found in 1966. Its production only began in 1990 after a large market was found for it in the United States. I doubt whether Schalz's criticism of Shell-BP on this score was well founded. However, that is not to endorse the company's recklessness when it finally began operation in Nigeria's Niger Delta.

Like Schatz, Rufus Akinmoladu was not only equally worried about the inability of the state to derive maximum benefits from its petroleum industry, but further argued that, the incapacity of the state to produce and refine crude oil independent of the oil transnationals and their service companies in the country were signs of its general underdevelopment. To Akinmoladu, while Nigeria had grown both economically and industrially particularly since the discovery of crude oil in the country in the late 1950s, it is still short-changed by the operating oil companies. In order to overcome this problem, he recommended that the Nigerian National Oil Corporation (NNOC), forerunner of the NNPC, be strengthened to enable the state have a firmer control of the oil industries.[3]

What Akinmoladu exactly meant by "maximum benefits" to the country is not really clear in his study. If he meant foreign exchange earnings from oil export and other fees, then, how will he explain the huge oil wealth that has accrued and still accrues to the Nigerian state since 1958, yet the well-being of the majority of Nigerians particularly those who live in the host oil areas is still worse than what it was before oil was found there. Even by 1976, when he conducted his study, the Nigerian state had huge oil revenue already accrued to it. Yet, rather than reduce, social and economic inequality has deepened since then. Suffice it to note that, more oil revenue accruing to the state may not necessarily mean the enhancement of the material well-being of the majority of Nigerians as long as state capitalism remains in tandem with that of the oil TNCs. In fact, this study demonstrates that, in spite of the huge proceeds accruing from the oil industryto the Nigerian state, deepening rural poverty still pervades the immediate oil areas in the Niger Delta. The NNOC was established in 1971 and abolished six years later, essentially because of its weakness; a weakness that mirrors to a large extent, the nature of the state Nigerian, its politics and mode of capitalist accumulation. If the NNOC must be strengthened in line with Akinmoladu's recommendation, it is doubtful if that will bring about any major change in the structure of dependence in the country's petroleum sector especially in its upstream sector. This is because nothing suggests in his kind of change that the state's approach to industrialization should undergo a rethink by placing emphasis on its capacity to produce locally, capital goods hitherto imported. For, all that is done at the moment is to reproduce some of them locally of substituting. Worst still, the country's iron and steel industries which ought to produce most of the oil tools locally that will help reduce its technological dependence on the oil TNCs, are ridden with crisis even at the take-off stage.

By the turn of the 1980s, a new body of literature in the Nigerian petroleum industry began to emerge. Scholars, this time, started to show concern on how the country's huge wealth accruing from the oil and gas sector was mismanaged by the state and its allies. Cases of large scale fraud in the country's petroleum industry such as the 2.8 billion Naira fraud was uncovered in 1979. The Irikefe-led Tribunal was set up in 1980 to investigate the fraud. Other issues that equally attracted scholarly interests at the time, were the neglect of agriculture since crude oil became the major foreign exchange earner; politics of revenue allocation; and neglect of the oil producing areas and protest by the people who live there. In virtually all the studies on Nigeria's oil industry during the decade of the 1980s, the dominant mode of analysis employed by most of the scholars was basically structuralist. This, in part, explained why they were generally descriptive in nature in spite of of the fact issues of underdevelopment and dependence were raised in the discourse. Let us begin with Odoliyi Lolomari, who himself, was an oil bureaucrat. He argued that, the colonial mining law, the Mineral Act of 1914, which was largely rephrased as the Petroleum Decree of 1969, was exploitative and vested absolute power over minerals on the state. And that the state's major aim of intervening into the country's petroleum sector was basically to shore up revenue for itself its allies and foreign capital. To stop all this, he called for review of the Mineral Act[4] On the domestic bourgeoisie, Lolomari was as critical as both Osoba and Ihonvbere. To him,

> The domestic bourgeoisie is an agent for the foreign oil companies either in the award of oil blocks or crude oil purchase by the independent oil companies. He is satisfied with being a silent partner waiting for dividends to slop and slop into the barrel of crude oil. He is an economic prostitute lacking in entrepreneural drive.[5]

Reviewing the Mineral Laws as Lolomari has suggested is quite in order. But the question is: How will it be done? Should it begin with the general re-orientation of the state, or the domestic bourgeoisie that he bounded as a prostitute?. He was silent on all this. It is important to underline the fact that, unless the state is itself, reoriented away from its current mode of capitalism to begin to indigenize technology, revising the Mineral Act will not necessarily bring about a reversal of underdevelopment of the country and its oil-bearing areas in the Niger Delta. Lolomari's analysis of the domestic bourgeoisie was really not different from that of Osoba and Ihonvbere in the sense that, he too, saw the domestic bourgeoisie in the oil industry as equally good for nothing. Like Ihonvbere, he created the impression that should it become fully involved in oil production, all will be well for the country. The earlier criticisms that I have levelled against Ihonvbere and Osoba in Chapter Two, are too clear to necessitate a repeat here. In other words, Lolomari's analysis did not really help matters much.

Particularly concerned with the role of the bureaucratic class in the 2.9 billion Naira scandal in the petroleum industry in which Shell, Mobil and Gulf were implicated, Saddique Mohammed accused it (the bureaucrats) for conniving with the oil companies to shortchange the country. To him, they are criminals, wasteful in the use of resources, and responsible for the lack of skilled manpower and technology in the oil sector.[6]

In her study on the state and oil technology in Nigeria, Terisa Turner has broadly classified the state bureaucrats into two groups namely: (i) the technocrats, who, equipped with a sound technological background, are bent in wresting oil technology from the oil tncs; and (ii) the general administrators, who, without any technical background, have favoured the reliance on the oil TNCs for the 'transfer' of oil technology to Nigeria[7], perhaps until such a time that the local technical skill is built up. Extending Turner's categorization to Mohammed's analysis of the bureaucrats, it is not too clear whether he meant the latter group, who at the time of his study, was really concerned with how much it could get as its own share from the oil wealth. It is not assumed that the former group is any different; for, as a class, it also helps propel the process of state capitalism. To characterize the state bureaucrats generally as comprador, however, bearing in mind Turner's analysis, as Mohammed did, does not provide any deep insight into the various factions of the state bureaucrats particularly how the struggle among and between them has come to partly explain the deepening dependence of the state on the foreign oil companies.

With the advent of crude oil in Nigeria and its export, the state has been awashed with huge wealth so much so that the problem was reportedly how to spend it. This was reflected in the manner the country's foreign relations was conducted which many scholars saw as playing a "father christian" role in Africa. Some scholars in the political science discipline became interested in investigating the linkages between Nigeria's oil wealth and its foreign policy. Among theme, was Kayode Soremekun, who studied the country's petroleum policy within the context of the federal structure of the Nigerian government using the input-output decision-making approach as his mental construct. He tried to establish how Nigeria, as an oil exporting country, was able to use its oil to pursue its external relations. He argued that the decision-making process of the government in the oil industry, gave much power to some influential civil servant citing how Philip Asiodu, who was then Permanent Secretary of the Ministry of Petroleum and Energy, and doubled as Chairman of the NNOC and the Petroleum Advisory Board, became the most powerful individual who decided the country's petroleum policy. What was later known as the country's petroleum policy, Soremekun contended, was indeed, his personal decision. He also contended that, the various joint venture agreements that the Nigerian government signed with the operating foreign oil companies, was informed by its (state) desire to shore up revenue. He rested his case with the submission that, though much revenue had accrued to the state, it did not remarkably influence its external relations.[8]

That Philip Asiodu, as a generalist (bureaucrat) and one of the super permanent secretaries under the Gowon's regime, and whose office was vested with much political power over petroleum, swayed the direction of the country's oil and gas industry is not open to dispute. For one thing, Asiodu only acted in protection of the state of which he was a part. Suffice it to note that, his action was personal as Soremekun's analysis showed. Concerned therefore, with how the bureaucratic class with largely compradoral interests could get a larger share of the oil wealth, it was not strange that he, Asiodu wanted the NNOC still managed as a parastatal, in spite of its woeful failure to build up the technological base for the nation's infant oil and gas sector. So, from 1971 when it was established until its dissolution in 1977, it was regarded as a gestation period, while bureaucratic capitalism lasted. The state's urge to consume the oil wealth without much concern for its generation crippled the NNPC which came on stream in 1977 as evidenced in the Corporation's failure to regulate the operation of the oil TNCs in the Niger Delta. The decision-making process which vested power on Asiodu may not be the major problem why the state of affairs in the country's petroleum industry was and is still in a flux as Soremekun wanted us to believe. Rather, the nature of state capitalism provides a more useful insight into it. In fact, state capitalism sheds more light on its ultimate aim of intervention into the country's petroleum industry through the joint venture agreement. For, experiences of its operation since 1972 show that joint venture was not used by the state to acquire the capacity to find, produce and refine crude oil independent of the oil companies as Algeria, an African OPEC member did in the 1970s. Rather, joint venture agreement was and is still operated to shore up more rent, leaving the control of oil technology where it has always been: with the oil TNCs. What Soremekun regarded as shortchanging of the Nigerian state by the oil TNCs, is not really so. For, that is what the Nigerian state model of capitalism guarantees Shell and other oil TNCs in order to reproduce the process of accumulation in the global petroleum industry. By so doing, its own basis of survival is also ensured, amid deepening dependence. Obviously, this is to the detriment of the well-being of the majority of Nigerians and the environment, but it does not matter to the state managers and other beneficiaries of the exploitation.

Nigeria's membership of OPEC has, no doubt, brought some gains to the country especially in terms of a better price regime. However, the OPEC meant more than just securing a good oil price for its members; it wanted them to have adequate control over their petroleum industry. Control that should translate into an indigenous technical capacity to find, produce, and refine crude oil independent of the oil transnationals. As earlier noted, Algeria, an

OPEC member, proved that this goal is achievable as the Algerian state took charge of its oil sector in the 1980s. Yet, in Soremekun's analysis, all that one could get on this, was a 'continuous members' of the OPEC because of what he described as the 'turn out of new events such as changes of government and personnel in the Petroleum Ministry'. Granted that he was cautious, what he meant by all this clauses was altogether not clear in his study. Worst still, he was so much concerned with the internal decision making process in the country's oil industry that he did not give real insight into the linkage between country's external relation and its oil wealth.

Last but not the least of the studies on the Nigerian petroleum industry by the structuralists, is that of Cyril Obi. In his investigation of the relationship between the Nigerian state and oil, Obi argued that, the struggle among the various actors/actresses and institution which make up the Nigerian state structure is merely for the political control over wealth generated from the oil and gas industry. Without developing any clear theoretical linkage between the country's petroleum industry and the IMF/World Bank-inspired Structural Adjustment Programme (SAP), much more discussing its impact on the latter, Obi contended that, adjustment had deepened the country's petroleum industry dependence on the oil TNCs. Furthermore, he argued that, the de-investment by the state in both the joint venture agreement with the oil transnationals, as well as in equity of three oil marketing companies, were meant to shore up revenue, and restructure the oil industry generally. He submitted by prescribing the creation of a revolutionary state that could effectively control the oil industry, and build key infrastructural amenities like roads, hospitals and portable water in the utterly neglected oil-bearing areas in the Niger Delta instead of using the local chiefs and contractors whom he indicted for non-performance and frauds.[9]

Like Turner, Obi limited political competition in Nigeria to that of grim struggle over the control of the oil wealth. That is rather unrigorous. In fact, in our earlier critic of Turner of, the danger of reducing political competition in Nigeria solely to the level of accumulation was already pointed out; so, too, was the need to broaden the scope of the country's politics emphasized. All has come to make it unnecessary to undertake a repeat of the criticism of this issue here. But that is not all about the misgivings in his study. At one level, Obi argued that the SAP project of the Bank and the Fund had deepened the state's dependence on the oil companies; yet, at another, it (SAP) enabled the state to shore up revenue. Not only did he fail to shed light on these claims, they are contradictory. For, the real issue at stake when the state adjusted in 1986, is not that SAP had furthered the reinsertion of the country's economy into the global capitalist system thereby weakening the state economically. How the state used SAP therefore, to shore up more revenue as Obi claimed was not clear. In fact, the SAP-induced de-investment in the Nigerian oil and gas industry signalled that, the state had bowed to the pressure of the Bank/Fund to privatize, and not that it was meant to raise fund. At the time of this study, the Nigerian state had no equity interests in any of the operating oil TNCs in the country. And that has not changed. So, no shares could have been sold in the joint venture as Obi wanted us to believe.

Lastly, Obi's recommendation that compensation previously paid to the chiefs in the oil communities in the Niger Delta should be stopped is in order. But he did not tell us who should be paid such compensation. I think his suggestion that the state should undertake the construction of social infrastructures like road, electricity in the oil areas may not really help matters since the state and the oil compaies adopt the same approach to the provision of social amenities in the oil areas. That is, to further their exploitation, but not to develop them. That is not all as the nature of the political economy of underdevelopment of the oil areas reveals a lot of in-fighting among the constituent groups of the state and some elites of the area of who gets what contracts at what cost. Contracts that are in most cases, abandoned half-way since they are largess from the state. Worst still, both the state and the oil companies regard the oil areas as minefields and must be exploited as such. Social amenities like road, is therefore, only a modality of exploitation and underdevelopment of the Niger Delta since its aim as an access road, is basically to link up oil locations in the minefield. For, soon after an oil well becomes extinct, both the state and oil companies abandon the road. Above all, the issues of compensation and community projects as implied in the joint venture agreement, are basically political in nature. Neither the Nigerian state nor Shell would ever want to own up to the environmental degradation of the host oil communities under the joint venture. For instance, Shell has regularly made it clear that it is just an operator of the agreement; and in that capacity, the company can not be said to be legally, to pollute its host oil areas. Therefore, it must not be held for compensation, nor for providing social amenities for the people of host oil areas. Instead, it is the joint venture that is the culprit. Even if it is so, which is not in anyway, the ratio of their participating shares and not *equity* in the agreement, dictates the amount to be borne by the partners, that is, the state and Shell as at the time of this study. This means that, the state represented by the NNPC, has to bear the lion share of 80 percent, leaving Shell with 20 percent. This, the state is not prepared to shoulder. Instead, it imposes fine for gas flaring. But whatever Shell pays as fine, is finally charged to the joint venture as production expenses, meaning that it is the state is the ultimate loses with its oil-bearing areas badly hit. Worst still, since state capitalism is in tandem with that of Shell, none of them is really willing to pay compensation for crops damaged, nor ready to sanitize the environment. All has made it difficult to understand how Obi's recommendation could be achieved. His 'revolutionary state' has hardly any prospect of emerging at the moment as the existing order of capitalist accumulation in the economy generally, and the petroleum industry in particular, is further entrenching the state and its constituent groups, and deepening rent-seeking.

Marking a significant departure from studies on the Nigerian petroleum industry by the structuralists, is a group of scholars who, employing the Marxian political economy approach attempts to shed light on the character of accumulation in the oil sector and the nature of underdevelopment it introduced in the Niger Delta and the country at large. Like the UDT, the theoretical inspirations of this group comes from Marxist-Lennist postulations. At the heart of the issues addressed, are the Nigerian state, and its mode of capitalism, and the political economy of the oil TNCs, and the underdevelopment of the country. The literature on Nigeria's petroleum industry from this perspective is still scanty. In fact, this study is partly an attempt to fill the gap in the field.

In a major work on the state, labour and class struggle in the Nigerian petroleum industry, Julius Ihonvbere argued that, the 'dependent rentier oil state' has, on account of its material interest in the country's oil and gas industry, connived with the oil transnationals to domesticate the oil workers in the name of 'industrial peace'. His study demonstrated how both oil capital and the state antagonized labour unions in the petroleum industry citing instances of how the former relied on threats, blackmail, bribery, victimization, manipulation, lies to divide and destroy trade unions, while the latter used its military to crush any labour action that disrupted oil exploration, production and transportation in the country. In spite of these anti-union measures, Ihonvbere's study further showed that, the oil workers remained more restive, devising various forms of protest to resist the high-handedness of both the state and the oil companies. He identified two forms of protest that were commonly used by the oil workers namely, overt and covert. Overt protest included general strike, blockade of company premises, holding management staff hostage and other non-violent measures such as sit-down and look. Covert protest, on the other hand, included deliberate lateness to work, delay in carrying out order, deliberate industrial accidents, cutting oil pipelines, theft, dumping of drilling equipment into the high seas etc.[10]

As protests become more deadly in the country's oil and gas sector, Ihonvbere questioned Norre *et al*'s thesis on 'aristocracy of labour' in the global oil industry which was propounded by Norre, Hein and Bird.[11] To him, the nature of labour union in Nigeria's petroleum industry refuted Norre's postulation on labour in oil 'industry'. In essence, Norre ought to have taken note of the nature of state capitalism particularly its alliance with the oil transnationals before reaching such conclusion. How the state and oil TNCs disorganize trade unions and set no standard for condition of service of the oil workers, and the consequences of all this for the fierce class struggle are too crucial to be neglected if one is to have fuller understanding of the relationship between state, foreign capital and labour in the petroleum industry.[12]

Drawing on reported cases of protests in both the up- and down-stream sectors of Nigeria's oil industry, especially on how petrol attendants, tool pushers, oil tanker drivers have, through various protests, paralyzed the oil sector, Ihonvbere reached the conclusion that, rather than an aristocracy of labour, oil workers are really prolatarianized. He recommended that labour should be adequately compensated instead of the use of force and punitive policies like the 1975 Anti-Sabotage Decree with the death penalty.[13]

Ihonvbere's argument is generally well taken, although some important issues omitted in his study that will help gain a fuller understanding of the negative impact of the operations of the oil TNCs on Nigeria's economy including the Niger Delta, need to be addressed. As shall be shown in Chapters Six and Seven and Eight of this study, majority of the oil workers, casual or permanent, with Shell and other oil TNCs, hail from the oil-producing areas in the Niger Delta. Notwithstanding the category, as oil workers, they are legally forbidden from agitating the constant pollution pf the environment of their own community, or any other one in the Delta region. So, they often take full advantage of most protests in the oil-bearing areas to covertly vent their anger with the oil TNCs by assisting in acts of sabotage like cutting oil pipes. At times, they provide confidential information on how to disrupt production, all in an attempt to punish the state and the oil companies.

Ihonvbere's characterization of the Nigerian state as rentier does not really give us a clue to what he actually meant. Like

Turner and Forrest, he probably could have meant that since the Nigerian state extract its surplus from the country's oil and gas industry by depending largely on royalties and rents from the oil companies, and proceeds from oil export, it is rentier. As noted earlier in Chapter Two, this is not the real problem; rather, our concern is the use into which the rents being collected is put since virtually all states collect rent in one form or the other. What is an 'oil state'? Could it be one that relies largely on the oil industry for its revenue as Ihonvbere's characterization would want us to believe? For, if it were so, how would he, Ihonvbere brand this same Nigerian state before the advent of crude oil when agriculture was the mainstay of the economy? To risk a holistic characterization of the Nigerian state as he did, is hazardous. This is all the more so as state model of capitalism differs from one sector of the economy to the other. For instance, while the state relies on the oil TNCs for revenue, it recycles part of the surplus into the steel sector where it has assumed the status of a landlord. Characterization of the Nigerian state therefore, must be done within a specific context if its understanding is to be deepened.

So far, our analysis of these studies have clarified basic unsettled theoretical issues in the political economy of Nigeria's petroleum industry. Let us proceed to examine how those theoretical issues are addressed in the specific studies on the Niger Delta; an area where the forces of state and transnational capitalism has shaped and continues to shape the nature and direction of its underdevelopment.

3.2. Oil and the Economy of Niger Delta

Marxian political inquiry demands that, an idea of the pre-capitalist social formation of the Niger Delta will help deepen our understanding of how oil capital, this time, Shell, in conjunction with the Nigerian state have actually contributed to the underdevelopment of the region. Therefore, our inquiry begins with the content analysis of the early works on the economic history of the Niger Delta, though they are not Marxian in approach.

In pioneering a study on the colonial history of the Niger Delta, Kingsley Dike argued that, the British in particular, and most other Western European trading companies generally, came to the Delta area for three basic reasons. First, was the search for raw materials to feed their metropolitan factories in Europe; second, was the increasing need for cheap and slave labour to work in their plantation farms in the British occupied territories in the West Indies; and third, to secure market outlets for their manufactured goods brought from Europe. To Dike, a detachment of the Royal Navy was used in subduing all forms of resistance across the land; a variant of British gun-boat diplomacy that helped accomplish the economic goals of the colonial trading companies whose mode of operation was to structure the insertion of the economy of the Delta into the western capitalist system.[14]

Drawing on the particular experience of the Eastern part of the Niger Delta with British colonial rule, Dike traced how the policies of the colonial state and actions of the British trading companies like the United African Company (UAC) led to the destruction of the traditional political institutions of the area, insisting that it became widespread as they penetrated into the hinterland from the coast. The "house system" around which the political administration of the area was anchored, was particularly badly hit. The 'Native Court System with all its artificialities, was to replace the traditional penal system and jury. The attempt by the British colonial state to amalgamate the Igbo and Ibibio people, Dike's study showed, was ostensibly for commercial purpose, noting how it disrupted the peace and stability of the hitherto small but autonomous city-states. To him, the climax of the resistance of the intruding British colonial trading companies by the people of the Eastern Delta culminated in the various wars they fought across the area, and with the resultant dethronement and deportation of King Jaja of Opobo.[15]

Dike might be right in claiming that the process of the integration of the Delta economy into the mainstream of British capitalist began with the mode of surplus extraction of the colonial trading companies. However, the nature and process of the integration was not really explained in his study. Granted that the importation of manufactured good from Europe by the British trading firms was detrimental to the economy of the Delta. The question that arises is: How did it impair the development of the domestic business class in the area? Again, how did it hinder or promote the process of capitalist development in the Delta? These issues were not really addressed prompting one to question therefore, the nature of underdevelopment of the area. The main contention is not that King Jaja waged several wars with the British trading companies over the control of the trade in palm oil; but in whose interest. For one thing, King Jaja and other kings in the colonial Delta were themselves, part and parcel of the colonial state and its model of capitalist exploitation. In essence, if he, Jaja resisted the British, he acted largely in defence of his selfish interest and not for the well-being of the people. As it shall soon be made clear in the course of the work, traditional rulers have, since the advent of crude oil in the Niger Delta, been in concert with Shell in bringing about the underdevelopment of the oil areas.

For the impact of British rule on the Western Niger Delta, Obaro Ikime has, in his study, traced the root cause of the discord between the Urhobos and Itsekiris kingdoms to the policies of the colonial state. He argued that, hitherto mutual trade relationship that existed between the Itsekiris, who live in the coastal areas, and the Urhobos on the hinterland, was disrupted since the advent of the British trading companies in the area. For the purpose of better price regime for palm oil, the foreign trading decided to deal directly with the Urhobos, who were the real producers, instead of the Itsekiris, who were basically middlemen. This was to undermine the Itsekiri traders most of whom were thrown out of business. With the threat to the monopoly the Itsekiri traders had over palm oil business, they had to put up stiff confrontations to resist direct trade between the Urhobos and British merchant capital. Confrontations that not only left the Itsekiris further subdued by the British naval contingent, but resulted in the dethronement and subsequent deportation of Chief Nana of Itsekiri, who, before then, was known as the 'merchant prince' of the Western Niger Delta. All this helped retard the growth and development of the economy of the Western Niger Delta.[16]

Ikime's claim that British colonial rule introduced discord into the hitherto mutual relationship between the Urhobos and Itsekiris, is not open disputes. But that is not enough reason to blame as he did, the British merchant capital for the former's decision to trade directly with the foreign trading companies. For, the Urhobos, as producers of palm produce, acted in line with the logic of mercantile capitalism, by bargaining for better returns for their palm products instead of being exploited by the Itsekiri merchants, who as middlemen, wanted to reap all the gains. Granted the conflicts between the Urhobos and Itsekiris had led to the backwardness of the Western Niger Delta, it is not clearly demonstrated in Ikime's analysis the nature of the underdevelopment.

Worried about the negative impact of the British trading companies on the economy and people of Nembe situated in the East of the Niger Delta, Joseph Alagoa's brought into the fore, how the Royal Navy subdued pockets of resistance from the area as it advanced the economic interest of colonial capitalism in Nigeria. His main contention was that, the disruption of the traditional, political and economic systems of the Nembe city-state by the forces of British imperialism, laid the basis for the impoverishment of its people during the colonial era and after.[17]

Alagoa's study was not significantly different from the works of Dike and Ikime on the history of British colonial rule in the Niger Delta in the sense that he, too, basically indicted the colonial state and the British trading companies for the exploitation and backwardness of the Nembe kingdom. However, to contend as Alagoa did, that British colonialism destroyed the traditional political institutions without giving insight into the complicity of local allies of merchant capital in the plot, nor the role of the traditional or warrant chiefs in sustaining the economic underdevelopment of the Nembe city-state, is to neglect the situation of the internal structure. This, no doubt, impaired a deeper understanding of the political economy of the underdevelopment of the Niger Delta.

On common string of thought runs through these early studies; that is, they all presented a historical account of the activities of British merchant capital in the Niger Delta. Insightful as these studies might seem, however, critical issues like the nature of the colonial state, especially how, its mode of capitalism in conjunction with merchant capital, had structured the economic backwardness of the Niger Delta were not really examined. For instance, none of these studies provided a rigorous basis for deepening our understanding of the main motive behind the slave revolt. Quite true that the trade in human being was condemned, its prosecution was not without the tacit support of most of the local chiefs and produce merchants. Again, the pre-colonial Niger Delta was not without capitalist exploitative tendencies bearing in mind that its political power structure and mode of production were not really egalitarian. Yet, no attempt was made to highlight this and other related issues in these studies. Nor was there an analysis as to their status before the advent of British colonial rule and after. These are some of the neglected issues that this study addresses in its attempt to provide a deeper understanding of the nature of the underdevelopment of the Niger Delta both before and since petroleum was found there.

Since Shell among other oil producing transnationals began exploration and exploitation of crude oil in the Niger Delta, the environment and economic activities of the host communities have been severely damaged. This gave rise to a new body of studies on the socio-economic and scientific impact of their operations on the oil-bearing communities. Some of the studies were mostly done by scholars in the Biological Sciences like Botany, Zoology, Biochemistry. In virtually all these studies, the scholars were concerned with the exact impact of spilled crude oil and gas flaring on the environment especially the soil, flora and fauna of the host oil areas. Generally, the studies demonstrated that, palm trees most of which were wild, that were within a radius of one kilometer of the flare sites, grew tall without bearing fruits before dying finally; and that vegetable crops like tomato and pumpkin were not grown with the same radius because of the desiccating effect of the flare on the soil. So, too, did the surrounding vegetation most of which were swampy mangrove forests turned into 'derived savanna as a result of gas in both day and night.. Spilled crude oil, it was also shown, polluted both surface and sub-surface water, killing virtually all aquatic organisms like fish and plankton.[18] These results were terminal in nature instead of constituting theses for further researches in the Social Sciences. References, therefore, are made to them in the course of the discussion either to enrich or clarify arguments on the underdevelopment of the Niger Delta.

Social Science-based studies on the Niger Delta since crude oil was discovered there, are still scanty. Let us examine them.

In their study, Eniola *et al*, have argued that, the Niger Delta oil-areas benefitted tremendously from the oil companies by way of using infrastructural amenities like roads provided by the oil TNCs, to enhance the economic well-being of the people who live there.[19] As noted earlier, access roads are really meant to facilitate exploitation of the oil bearing areas but to develop them. If the oil producing communities situated close to these amenities ever benefitted from some of them therefore, it is by sheer coincidence. For instance, all the roads in the Ughelli oil areas are Shell's access roads connecting the company's various oil locations there. And as long as the company still operates there, the roads will be maintained, but are abandoned when its oil wells dry up. In essence, as soon as Shell has no producing well in the area, it pulls out leaving the roads to fall into disuse. Worst still, the state, does not undertake the maintenance of such roads. Suffice it to note that, what has passed for Eniola and others as development, is really nothing but a mode of capitalist exploitation adopted by the oil TNCs including Shell to facilitate accumulation in Niger Delta. This is all the more so since these so-called amenities do not really help uplift the material well-being of the people who live in the immediate oil-producing areas.

In both the up- and down-stream sectors of Nigeria's oil industry, land is always lost to both the state and the oil transnationals. With the relative recklessness with which the oil companies operate in the Niger Delt, land crisis is worsening in a predominantly farming community. All has led to conflicts within oil areas, and between them, the state and the oil TNCs. Concerned therefore, with the impact of the Land Use Decree of 1978 on the oil areas, Obioma argued that, it was a plot by the state to vest sole ownership of land in itself, and as well, protect the oil companies from land litigations in the oil areas. As the decree vested the state with absolute ownership and control over land where oil is explored, produced, transported and stored, Obioma further contended that, the people of the oil-producing areas were turned into squatters in their own ancestral home.[20]

Obioma's position on the Land Use Decree is well taken. But it is important to add that, in terms of payment of rents to landldords, the decree did not apply to land where the oil TNCs had been operating before 1978. So, on which they still pay rents to the landlords in the oil areas. Again, though the state claims to pay compensation in spite of the Decree, it is not for land, but crops, and at ridiculously low rates that were fixed by the state and the oil companies. Details on this issue will be taken up in subsequent Chapters.

This Chapter ends with an analysis of the study by Eboe Hutchfull on the impact of oil-pollution on the environment of the Niger Delta. Hutchfull's main contention was that, the problem of oil pollution in Nigeria's oil-bearing areas had been exacerbated due to the absence of effective state policies on the environment and the predatory attitudes of the oil companies. Drawing on the specific case of Finiwa - 5 Blowout, which occurred on 17 January, 1980, Hutchfull demonstrated that, despite the fatal destruction of aquatic life by spilled crude oil from the rigs of Chevron Oil Company and Texaco Overseas Petroleum Company, the state defended them, claiming that no serious damage was done to the ecology of the area. Hutchfull submitted that, for the oil companies to be cautious in their operation especially in protecting the environment, the state must itself, be serious with checking oil pollution.[21]

Hutchfull's contention that the Nigerian state is weak and has no effective policies on environmental protection resulting in the recklessness of the oil companies, is well received. It is important to add that, its (state) weakness stems also from the adoptor of Oil in Navigable Waters Act No.34 of 1968 among other state laws on the oil and gas sector, from the corpus of international laws on the environment without the capacity and provisions detailing out the necessary ingredients of the subject matter. Furthermore, I think, the issue at stake is not so much the weakness of the Nigerian state as its impact on the oil areas. It was expected that Hutchful would have touched on this in order to have a fuller understanding of the nature of the underdevelopment of the area. I think also that an idea of the nature of state capitalism in the oil sector, which was lacking in his study, would have shed more light on the state alliance with the oil TNCs and its implications for the environment of the Niger Delta.

From our critical analysis of these studies thus far, a more rigorous basis for gaining a deeper understanding of the nature of underdevelopment of the Niger Delta has been created. However, certain basic theoretical issues that are thrown up in the discussion still need to be contextualized in order to enhance our analysis of how Shell and the state pillaged the Niger Delta.

3.3.Contextualizing Some of the Basic Concepts Used in this Study

In this Section, therefore, our goal is to put into perspective, other basic concepts as applied here, having explained earlier those of underdevelopment and dependence in the Chapter One. In doing this, cognizance, is taken of the dynamics of change taking place in both the global petroleum industry and that of Nigeria, and the implication of this for the environment of the Niger Delta.

Capitalism, as used in this study, means that system of commodity production and exchange based on the private ownership and control of the means of production and the exploitation of wage labour and the peasantry by capital whether private or public. Under this broad definition, it is assumed that expansion and not stagnation, is the primary goal of capitalism and that the process of that expansion, however, also engenders stagnatory effects on a given social formation as evident in the activities of Shell in the activities of Shell in the Niger Delta.[22] Suffice it to note that, this study locates state-Shell capitalism in the country's petroleum industry within the twin logic of capitalist accumulation: expansion and stagnation.

The State: As shown in the above studies, the debate on the Nigerian state is not only continuing, but also in a state of flux. On its nature, while Claude Ake still regards the Nigerian state as 'colonial' in character, Tom Forrest brands it 'rentier'. Yet, Osoba and Ihonvbere characterize it as 'parasitic'. In essence, there is no holistic characterization of the Nigerian state. Even any attempt to risk such exercise outside of any context will not help matters. But, as it remains central to the process of capitalist accumulation in the country, a characterization that will enhance our understanding of the political economy of the Niger Delta's underdevelopment must be sought. There is need therefore, to look into the environment of global capitalist accumulation which is often neglected in previous attempts at characterizing the Nigerian state. This is all the more so as international monopoly in oil capital has been spreading its imperialism through the state in the world petroleum industry since the 1960s. In fact, since then, the state has often be regarded as a project of transnational capitalism.

In this study, therefore, it is recognized that the post-colonial Nigerian state does not behave in a significantly different way from the advanced capitalist state in the North especially since the era of the imperialism of international finance capital. This is because the Nigerian state acts to reproduce the aims of finance capital, while also ensuring its own means of accumulation in the process. Thus, both the state and capital have become part and parcel of the same global process of capitalist accumulation, performing basically the same role in upholding the international capitalist system. The reality is that, the Nigerian state, acts to protect the interest of international finance capital originating from the United States, Britain, Japan, France and West Germany in addition to that of indigenous (Nigerian) capital. So, too, the British state has, within its geographical confines, protected the interest of capital from all over the world by operating side by side and together with indigenous British capital. In fulfilling the role of providing the domestic conditions congenial to capitalist development and accumulation, both the Nigerian and British states, as with all other capitalist states either in the South or North, cannot be said to be defending and protecting just their specific national capital, but also, in one and the same process, upholding the interests of other factions of international finance capital of which indigenous capital and the state itself, are an inseparable part.

If as explained above that the post-colonial and advanced capitalist states both exist in the era of the imperialism of international finance capital to cater for their basic selfish interests, as well as those of indigenous capital, then the *Nigerian state* in the context it is used here, is characterized as *transnational.*[23] It is not to assume however, that, as a transnational state, it has equal capacity like its counterparts in Britain or USA. Rather, its technological and economic weaknesses vis-a-vis the strength of the advanced capitalist state are recognized. Therefore, the Nigerian state is not expected to effectively tackle environmental problems like the British state does. Finally, on its meaning, which is also still in a condition of flux. While recognizing the various institutional groups like the domestic bourgeoisie, the traditional ruler, the military, and local capital all of which make the state a reality[24], the nature of struggle between and among them for accumulation make it (state) not to be as regarded as a public force which is objective in the sense that it uses its monopoly of coercion to police and guide the society rather impartially since it is not class neutral. Rather, the state as used here, is a specific modality of class domination which autonomizes the system of institutional mechanism in order to dominate society.[25]

Class is used in this study to mean a group of people, who differs from each other by the place they occupy in a historically determined system of social production relation (in most cases fixed and formulated in law) to the means of production, and to the state, role in the social organization of labour, and the dimensions of the share of social wealth of which they dispose and the mode of acquiring it.[26] Within this broad context, it is noted that though both the state and Shell constitute the capitalist class in Nigeria's petroleum industry since together, they own the means of production, the latter has used its control of oil technology to subordinate the former. In essence, Shell is not only exploiting the state but using it to underdevelop its Niger Delta in the process of capitalist accumulation.

One of our major contentions in this study is that, Shell only came to conclude the process of the underdevelopment of the Niger Delta that began in 1900 when the British imposed colonial rule on the area. In fact, Marxian political economy demands that it is important to have an idea of how the imposition of British colonial rule had laid the foundation for the capitalist exploitation of the Niger Delta is crucial if the role of Shell in the pillage of the area is to be fully understood. It is to this that the next Chapter now turn.

Notes and References

1. See Scott Pearson, *Petroleum and the Nigerian Economy*, (California: Stanford University Press) 1970.
2. Ludwig Schalz, *Petroleum in Nigeria*, (Ibadan: Oxford University Press) 1968.
3. Rufus Akinmoladun, 'Oil in Nigeria: A Study in Political Economy of Development', Unpublished Ph.D. Thesis, Political Science, Howard University, Washington D.C. 1976.
4. Odoliyi Lolomari, 'The Evolution of Nigerian Oil Policy' in *Oil and the New International Economic Order, Proceedings of the 1976 Annual Conference of the Nigerian Economic Society*, Nigerian Economy Society, Ibadan, 1979.
5. *ibid*, p.23.
6. Siddique Mohammed, 'Petroleum and Energy' in M. Kayode and Yusuf Usman eds., *Nigerian Since Independence Vol.II*, (Ibadan: Heineman) 1969.
7. Terisa Turner *op. cit.* (1976).

8. Kayode Soremekun, 'Nigerian Petroleum Policy and Nigerian External Relations, Unpublished Ph.D. Thesis, Department of International Relations, University of Ife, Ile-Ife, 1984.
9. Cyril Obi, 'Oil and State in Nigeria', Unpublished Ph.D. Thesis, Department of Political Science, University of Lagos, 1992.
10. Julius Ihonvbere, 'Labour, State and Capital in Nigeria's Oil Industry', (forthcoming book); See also his 'Class Struggle in the Oil Industry: The 'Rentier' State and Labour Control in Nigeria' (mimeo) and 'Resistance and Hidden Forms of Protest Amongst the Petroleum Proletariat in Nigeria' (mimeo).
11. Peter Noore, *et al.*
12. Ihonvbere, *op. cit.*
13. *ibid.*
14. See Kingsley Dike, *Trade and Politics in the Niger Delta ...* (Oxford: Carendon Press) 1986; see also, James Anene, *Southern Nigeria in Transition: 1885-1906*, (Cambridge: University Press) 1966.
15. *ibid* (Dike).
16. Obaro Ikime, *Niger Delta Rivalry: Itsekiri-Urhobo Relations and European Presence, 1884 - 1936* (London: Longman) 1969.
17. See Joseph Alagoa, *The Small Brave City-States* (Ibadan, Ibadan University Press) 1964.
18. For notable scientific works done on the environmental imbalance caused by the activities of the oil companies, Anthony Imevbore, 'Studies on Toxicity of Some Nigerian Crude Oils to Some Aquatic Organisms (mimeo) 1989; L.C. Amajor, 'The Ejamah-Ebubu Oil Spill of 1970: A Case History of a 14-year Oil Spill'; I. Ekweozor and R. Snowden, 'The Studies of the Impact of a Minor Oil Spillage in the Upper Bonny Estuary', in (Lagos: *The Nigerian Petroleum Industry...* NNPC/FMWH) 1985.
19. See O. Eniola *et. al.* 'Environmental and Socio-Economic Impacts of Oil-Spillage in the Petroleum Producing Riverine Areas of Nigeria', (mimeo) 1982.
20. See B. Obioma, 'The Oil Companies and the Oil Communities in Nigeria: An Unequal Relationship', (mimeo) 1986.
21. Eboe Hutchfull, 'Oil Companies and Environmental Pollution in Nigeria' in Claude Ake ed., *Political Economy of Nigeria*, (London: Longman, 1985).
22. Abstracted from Karl Marx, *Capital Vols.I, II land III*, (Moscow: Progress Publishers) 1954, 1956 and 1965.
23. For details on the state in Africa, see the various works of Mamdani, Shivji, Alavi, Bangura, Beckman, but in particular, Adebayo

Olukoshi's 'The Post-Colonial State in Africa: A Debate Revisited' (mimeo) 1991.

24. Ralph Miliband, *The State in Capitalist Society*, (London: Weidenfel and Nicolson) 1969.
25. Claude Ake, *Political Economy of Nigeria op. cit.*
26. Abstracted from M.I. Volkov ed. *A Dictionary of Political Economy* (Moscow, Progress Publishers), 1981, p.43; Karl Marx *op. cit* K. Marx and F. Engels, *The German Ideology: Collected Works*, (London) 1976; and Marx Weber, *Theory of Social and Economic Organization*, (Cambridge: University Press) 1966.

Chapter 4

British Rule and Underdevelopment of Niger Delta

4.1. Introduction

The British government 'formally colonized' Nigeria including the Niger Delta in 1900 after the Charter it granted to the Royal Niger Company was revoked in 1899.[1] By 1937 when Shell first arrived in the Niger Delta to prospect for crude oil therefore, the colonial government had, through its policies and actions, as well as the activities of British trading companies, overturned its development continuum, and then put in place modalities for capitalist exploitation and accumulation all of which structured the economy of the Delta area. It is pertinent to first understand the operations of colonial capitalism in order to properly situate the role of both Shell and the Nigerian state in the underdevelopment of the Niger Delta. This chapter therefore, first examines the organization of social, economic and political activities, taking special cognizance of the mode of production in the pre-capitalist period of the Niger Delta; that is, before the imposition of British colonial rule. Second, it critically analyses the import of the policies and actions of British rule on the Delta area, noting the structures of capitalist accumulation that were installed to ease the exploitation of the area by foreign capital including Shell.

Before the discussion proper, it is important to explain dates as they are used in this study. The choice of 1900 as marking the beginning of British rule in the Delta is by no means an assumption that the British were the first among the European merchants to come to the Niger Delta. It is recognize that the Portuguese slave merchants had even distorted the development continuum of the Delta region before the British. Before 1900 too, the monetization of the Delta economy was still incomplete because of the prevalence of commodity currencies. This was all the more so because there was no universally accepted currency in the Niger Delta, which, in turn, limited financial and banking operations not only in the Delta area, but across the colony.[2] Furthermore, powerful as the

Royal Niger Company (RNC) was in the economy of the Colony inclusive of the Niger Delta, it still could not singularly withstand the stiff competition posed by other merchant capitals from Europe other than those from Britain. Worst still, the RNC itself, became pretty difficult for the British government to control. These and other related factors compelled the British government to revoke the charter granted to the RNC and in 1990, formally assumed the administration of a territory that later became known as Nigeria in 1914.[3] It is equally crucial to note that, though the colonization of Nigeria did not occur at a different time from that of the Niger Delta, the features of pre-colonial Nigeria varied from one region to the other. While noting the features of the pre-colonial Nigeria it is those unique to the economy of the Niger Delta, which were overthrown by the forces of colonial capitalism that are discussed here. Now back to the real discussion, beginning with the pre-colonial period.

4.2. Pre-Capitalist Economy of the Niger Delta

To begin with, the pre-capitalist period of the Niger Delta is used here to also mean its pre-colonial era. This is not to assume however, that there were no traces of capitalist mode of production in the Delta region during this period since its society was not egalitarian. No doubt, they existed. Rather, they were rudimentary, making the economy of the Delta area pretty difficult to be described as capitalist.

Understanding the pre-capitalist social formation of the Niger Delta or that of any other society for that matter, provides some insight into how its contemporary came to be. That is, the way in which the present-day phenomenon in the Niger Delta area are largely the results of the distortion, and re-orientation of its non-capitalist mode of production by foreign capital its selfish political and economic interest through the imposition of British rule. Suffice to say that, it is only with the knowledge of the pre-capitalist economy of the Niger Delta that a deep insight can be gained into how its development continuum was disrupted by the forces of Western imperialism under the aegis of British rule, structured its present realities. One of the main features of the pre-capitalist economy of the Niger Delta is its mode of production. This is because it enhances our understanding of how hitherto 'non-exploitative' system of production in the Niger Delta was relatively re-organized and re-oriented by foreign capital to promote capitalist accumulation with the advent of British rule. Non-exploitative as used here does not necessarily mean that the Niger Delta society was qualitative as used here does not necessarily mean that the Niger Delta society was egalitarian as earlier noted. On the contrary, it means that, though exploitation was part and parcel of the its production system, it was not of the scale that came to dominate the area as colonial capitalism took root.[5] More details on these and other issues are taken up in the course of this chapter. If it is agreed that the pre-colonial economy of the Niger Delta can hardly be regarded as egalitarian, it is order to examine the nature of its polity.

4.2.1 Political Structure

As Dike, Aloaga and others have clearly demonstrated in their respective studies on the pre-colonized Niger Delta, the area was made up of nations which were politically organized almost in the manner as they Greek city-states.[6] This was irrespective of the fact that while some of the nations were monarchical in structure, others were republican. Among the monarchies, was the Itsekiri Kingdom where the king as head of state, was nominated from the ruling house. Which means ascension to the throne was hereditary; a process that bestowed privileges and wealth upon the ruling class. Most of the nations that were monarchical in structure were believed to be people who were descendants of the Oduduwa.[7]

Unlike in the monarchies where the throne was hereditary, the societies, nations that were republican in structure, age grade was a major determinant of who was crowned as the traditional head of the area. Which means that in most cases, elderly men were usually at the helm of political affairs. Age also played dominant role in choosing who became a chief, a class of people who played prominent political role in the Niger Delta. Because of the non-hereditary nature of the traditional head of nations that were republican, an incumbent was removed without much ado. This was the exact opposite in the monarchies where the removal of the incumbent had to follow prolonged procedures, customs and rites, which often times, were cumbersome. Because of this, an unpopular king was caused to commit suicide or banished from the land.[8]

Nations across the Niger Delta, whether monarchical or republican in nature, maintained an army ostensibly for though not of a standing nature was maintain. It was usually raised from the age grades like the 'elotus'[9] otherwise known as the warriors in case of the Urhobo, or the 'canoe and war houses' in case of the Bonny monarchy[10]. The army was used ostensibly to maintain both political stability and to repel external incursions. This is by no means, an assumption that inter- and intra-ethnic wars were not fought particularly over farm land and fishing waters during which the army was even deployed to engage in the war. There existed such cases. For instance, the war between the Itsekiri and the Agbon people in the hinterland of Urhobo nation in the 19th century was one of such wars.[11]

Although holders of traditional political offices such chiefs enjoyed certain privileges in both the nations that were either anarchical or republican they were benefits that did not strictly accrue to them from their position on the production system as was the case in the colonial period and after.

Rather, kings, chiefs and elderly men were relatively wealthy by virtue of their traditional and political positions in the society. Some of them also engaged in farming among other economic activities, and became relatively wealthier than the majority of the people who were not titled. All this made inequality inevitable in the pre-colonial system, but its scale could not be compared with that in the colonial period when the forces of Western imperialism had taken roots. A look at the mode of production in the pre-colonial economy is in order at this point.

4.2.2 Mode of Production

Because of its centrality to the process of accumulation, an understanding of the mode of production in pre-colonial Niger Delta will assist us in capturing in greater details, the nature and process of its underdevelopment which was engendered by colonial capitalism. In discussing the pre-colonial mode of production in the Delta area therefore, our main concern in this section is to examine how its pre-capitalist production system was to reproduce itself, noting its basic features.[12]

To start with, though land, labour, capital and entrepreneur were the basic factors of production in the pre-colonial Niger Delta, land and labour were most commonly used. This does not mean the other factors were redundant. Rather, capital in the sense of industrial and monopoly capitals whose major aim is to have control over other factors of production had not been really developed to such scale. However, it still constituted part of the unconsumed wealth used to generate more wealth. Like capital, entrepreneurship was of limited scale, largely based on single proprietorship or family units dominant means of production, was communally-owned.[13]

As agrculture was the basic economic activity of the people, agriculture played a prominent role in the production system. How much agriculture was able to serve as the lifeline of the pre-colonial economy was contingent upon the nature of land ownership system and its pattern of usage. Across the Niger Delta region, land was communally owned, though the elders held it in trust for other members of the family or community. Individuals and families also owned land that was allocated to them from the community land. Non-members of the family or migrant farmers from other villages/communities were also allowed to obtain land for the period of cultivation without being charged a rent. Where rent was paid , which was the exception rather than the rule, it was usually in the form of a fraction of the harvest made during the season. It was common among the Urhobos, and others who lived in the upland of the Delta region. So, too, in the coastal areas where fishing waters were also communally-owned, members of other families or migrant fishermen obtained permission to catch fish in such waters. As in the case of land, rent, if charged, was usually a fraction of the total catch made.[14] Under the pre-colonial tenurial system, therefore, it was difficult for an individual or family to sell land. Although there were few communities among the Urhobos and the Ekweres who lived in the hinterlands of the West and East Niger Delta respectively where land was leased, it was the exception rather than the rule.[15] Details on land matters are treated in greater details in the course of the discussion.

Given the scale of production in the pre-capitalist economy of the Niger Delta, its essence was to satisfy the immediate needs of the family, though minimal surplus was accumulated to meet other social requirements. But the scale of the surplus created could hardly be compared with that arose from transnational capitalism when colonialism installed.[16]

Concerned with the search of means of survival, labour, most of which was unskilled, moved both between and within the various nations in the Delta area. However, on account of the small scope of production in the pre-colonial economy of the areas, labour movement was not engendered by the complete control of other factors of production by capital as was the case since the advent of colonialism, and in particular the rise of monopoly capitalism.[17] Rather, as land remained the relatively dominant factor of production, what took place was for labour to migrate seasonably from one community to another to work on it (land) for use value; thereafter, it returned to its home base. Agricultural labourers therefore, constituted most of the migrant labour across the Niger Delta in the period before British rule, though medicine men and craftsmen like blacksmiths also moved to communities other than theirs to sell their services often in return for agricultural produce and other necessities. This category of vocational migrant labour was common among the Akwas who lived in the hinterlands of the East Niger Delta.[18]

The character of migrant labour in the Western part of the Niger Delta was slightly different. Unlike in the East where medicine men and blacksmiths dominated the migrant labourers, those in West were drawn largely from the agricultural sector since most of the people were predominantly farmers. Unlike the blacksmiths and medicine men, migrant farmers produced most of the foodstuffs like palm oil/kernels from the communally-owned palm trees. Rent was paid in the produce and the remainder was sold to meet other social needs. The Urhobos and Isokos dominated the migrant palm tappers in the Western Niger Delta.[19]

Slave labour was not uncommon across the Niger Delta[20] In fact, it was the cooperation from the slaves that made labour organization relatively more efficient in the pre-colonial Niger Delta particularly in the case of general trade. It was slave labour also that was largely used to carry out the palm oil trade in the colonial economy of the Delta area so much so that great merchants of the Niger Delta like Chief Nana of Itsekiri, relied heavily on his slaves for accumulation.[21] The significance of slave labour was borne out of the geography of the Niger Delta itself. Geography, for instance, made the pre-colonial communities in the Delta region more or less dependent on each other for their socio-economic survival. For instance, while the people who lived in the hinterlands did not have the basic expertise and the raw materials for the production of salt among others basic needs, but was produced by those in the coastal areas, those who inhabited in coastal belt exchanged salt for foodstuffs like yam, pepper and palm oil from the former, which they could grow. That is not all as great merchants of the Niger Delta had, on account of their salt among others basic needs, but was produced by those in the coastal areas, those who inhabited in the coastal belt exchanged salt for foodstuffs like yam, pepper and palm oil from the former, which they could grow. That is not all as great merchants of the Niger Delta had, on account of their early contacts with the Portuguese slave traders, acquired manufactured goods such as bangles and other jewels made of gold, copper and silver. These social valuables which were regarded as a mark of wealth among the people, were later used as commodity currencies.[22]

Trade and exchange, therefore, became so central to the capacity of the pre-capitalist economy of the Niger Delta to reproduce itself. And the success of this was attributed to the extensive use of slave labour. Notwithstanding the defects of slave labour, it helped prosecute trade across the Delta area and brought the different kingdoms and people of the area relatively much closer.[23] Slave labour was still illegally used in the Niger Delta long after its abolition in 1809[24]. Part of the reason for this, was that some of the slaves were so hardworking that their masters could not have succeeded in their businesses if they were freed.[25]

Although, trade in the Niger Delta blossomed largely through the ownership of the means of production was still vested in the slave owners. For instance, Chief Nana of Itsekiri, as a merchant prince, owned the canoes (the implements of labour), the slaves (the objects of labour), and land (factor of production). The same was true of other forms of production like farming. Under such circumstance, production could not but be geared to meet the parasitic requirements of the slave masters. Though the slaves were made to believe that they were part of the entire family, they were not free men. Thus, they were paid in most cases, in kind; and labour was also forced. As a result, the slave who was society's chief productive force in the period before the advent of British rule in the Niger Delta, later realized he really had no stake in raising output of production. To the slave owner, the attitude of the slave was nothing but a contempt for work, believing that it is an occupation unfit for freemen, which, in turn, led him (slave master) to live a parasitic way of life. As slave labour began to make less economic sense since the slaves now wanted to use their own implements to work in their master's farms, the labour-rent system arose.[26]

Under the labour-rent system, the slave labourer would want to use his own implements of labour to work a part of the week on the land-owner's farm and the rest for himself on his own farm particularly as forced labour encouraged low productivity. For, the slave labourer became really more interested in raising labour productivity and output only when he worked on his own farm than his master.[27] And this production relation did not interest merchant capital which the British was to introduce into the area under colonial rule. This was because declining productivity meant decreasing supply of raw materials to its metropolitan factories. The subtle rebellion of slave labour against landlords and capital was to be crushed later by British merchant capital. By the end of the 19th Century, one of the major pre-occupations of the British was to ensure that both land and labour in the pre-capitalist Delta area would virtually lose their dominance as major factors of production if colonial capitalism was to thrive. Thus,by 1900 when the British formally colonized Nigeria inclusive of the Niger Delta, the colonial state, together with the forces of western imperialism socialized the production systems. Factors of production like land and labour were later overtaken by British capital, which forced labour to sell its skill in order to survive. It is in order at this point to examine how colonial capitalism laid the basis for the capitalist exploitation of the Niger Delta by foreign capital.

4.3 The Colonial Period: British Rule and Capitalist Exploitation of The Niger Delta

This section examines the process of the incorporation of the Niger Delta area into the British capitalist system, noting its overall impact on its political economy. Specifically, it is concerned with the analysis of the policies of the colonial state and the infrastructures that it had put in place to facilitate the capitalist exploitation of the area by transnational capital not least Shell. But before doing so, let us clarify certain unsettled issue on the colonial economy of the Delta region.

As noted, earlier the formal colonization of the Niger Delta took place at the same period with the rest part of Nigeria.[28] However, it is important to note that the geography of the Niger Delta and its economy slightly made the impact of colonial capitalism on the area different from that on the communities in the northern part of Nigeria. For instance, the geography of the Delta area dictated to a large extent, the modalities of exploitation that were installed by the forces of colonial capitalism and the state. Since the Niger Delta was and is still crisis-crossed by rivers, colonial capitalism simply developed these natural routes of transportation to ease the movement of palm produce, a major exports of the country then, to the sea ports for onward shipment to Europe. Imports were also brought in through these routes into the hinterlands. The situation in Kano was different in the sense that railway lines were constructed to facilitate the movement of groundnut, another major export at the time, from the North to the coastal ports of Lagos and Port Harcourt from where they were shipped to the United Kingdom. Which means that, although the policy of the colonial transportation system was to evacuate exports and imports and not to really develop the colony, its impact on the inability of the various nations that made up the colony, to reproduce themselves socio-economically was of different magnitude. As a coastal area, the Delta region was particularly badly hit since it disrupted the natural fishing methods of the people among others. This was not the case in Kano, and other parts of the northern Nigeria where the construction of railway lines, across farmlands did not significantly hinder farming. In fact, the delicate ecology of the Delta area made it easily vulnerable to disruption. These and other salient features of the colonial economy of the Niger Delta are particularly addressed while discussing the structures for capitalist exploitation which colonial capitalism had put in place. Now back to the real discussion.

4.3.1 The Colonial State

If the British government vested the Royal Niger Company with both political and economic power that would enable the company to safeguard its economic interests in the Niger Delta, in the period before 1900, the need for a colonial state to perform the same, if not more powerful role by 1900, was imperative. Therefore, the colonial state was created ostensibly to perform the task it was best known for: to establish good and orderly government in order to make it easier for the foreign companies to accumulate and accumulate.[29] Which was why the amalgamation of the Northern and Southern Nigeria in 1914 was essentially to centralize the administration of the colony to facilitate its exploitation including its Delta area by foreign capital[30]. Little wonder why the only bond of political unity between the colony and any of its part like the Delta area and the colonial state after 1914, was in the person of Sir Frederick Lugard since the Delta economy was still left as purely a palm oil producing one and consumer of the products made from it[31]. Suffice it to note that, amalgamation only enabled Lord Lugard to preside over the economy of the Delta area particularly the financial proceeds that accrued from the export of palm produce among other exports from the area without the development of the people and communities of its people[32].

In essence, what the colonial state did in keeping 'good and orderly' government, was nothing less than the creation of a conducive atmosphere for exploitation of the colony including the Delta area. To enable the colonial state accomplish its basic aims, its next task was to ensure peace in the colony. This, it did through the enactment of various colonial policies such as the Minerals Ordinance of 1912 in case of the mineral sector. By the provision of the Mineral Act, the sole rights of ownership over minerals was placed on the Crown, her Royal Majesty. This also to defined the content of land ownership as well[33] As a consequence, minerals such as sharp white sand found in Ughelli area and lignite deposits at Okpanam and Ibusa all in the present-day Delta state of Nigeria were owned by the colonial state.[34] It was the same 1912 Mineral Act as amended in 1913 ad 1914, that remained in force until Shell came to the Delta area in 1937 to prospect for crude oil. And when crude oil was later found in the Niger Delta, its sole ownership was also rested in the state.

After the enactment of policies, the colonial state still needed to ensure that its legitimacy and political authority were fully established in the colony. It (colonial state) therefore, continued with the direct rule that the RNC had already put in place in the

Delta area before its charter was repealed in 1899 as against the indirect method of administration adopted in the northern part of the colony. Though of limited success, the colonial state used direct rule to gain the support of some traditional elites who were appointed merchant chiefs. Warrant chief were no less than stooges of the colonial state[35]. In fact, the duties of the warrant chiefs included assisting the colonial resident officers in the administration of the Niger Delta in particular and the colony generally. By so doing, they became important tools of colonial administration as they were charged with the settlement of debt, communal disputes, marriage, bride price, enlistment of adults for taxation and conscription labour. By virtue of their relatively privileged position, most of the warrant chiefs became rich through corrupt practices.[36] That was not all as the colonial state pitted the warrant chiefs against their people majority of whom were opposed to its exploitative policies. For carrying out the colonial policies, most warrant chiefs and their relatives were given administrative positions in government and trading companies mainly as clerks, interpreters, gardeners and other menial jobs.[37]

The colonial state therefore, established its authority and legitimacy in the Niger Delta through the combined efforts of its instruments of coercion and the warrant chiefs, and others who acquiesced to its policies and actions. It was from this group of local collaborators created by the colonial state however, that arose the social class notably domestic bourgeoisies such as Don Domingo, Prest, Okotie-Eboh and Mowoe. There were to become some of the early politicians in the West of the Niger Delta. They had their counterpart in the eastern part of the Delta areas. The same was true in other parts of Nigeria. However, the irony of it was that the origin of the agitation for the demise of the colonial state in the Delta area and Nigeria generally was championed by the same group of local collaborators. This was just an aspect of the contradictions of the colonial state. That is, the colonial state unconsciously created a political class that finally sacked it in 1960. But it really did not have any choice. However, the real issue is not in its demise but the aim of those who took over political power. Concerned with inheriting but not changing the policies and structures of exploitation of the colonial state, the custodians of the post-colonial state merely collaborated with foreign capital to further their selfish economic and social interests. Once the colonial state was firmly rooted, the rest task became a lot easier. Let us now dwell on other policies and structures of exploitation.

4.3.2 Monetization

As earlier explained, trading was one major feature of the pre-capitalist economy of the Niger Delta, with yams, salt, pepper and, fish serving as commodity currencies. These commodity currencies were still used side by side with those introduced by the Portuguese merchants such as dinar, gold and manillas uptill 1900 when the British colonized Nigeria. However, the physical nature of the commodity-currencies made them inconvenient to handle. Worst still, they were not convertible. Given these and other limitations, granting of large credit was highly minimized explaining why the pre-colonial economy of the Niger Delta was monetized.

With the imposition of British rule, commodity currencies were gradually abolished. The abolition was informed by the basic interest of colonial capitalism, which was to ensure the full integration of the economy of Niger Delta and that of Nigeria generally, into the British capitalist system. In essence, monetization further incorporated the economy of the Delta area into the British capitalist system, by facilitating the pervasive use of the British Pound Sterling (£), its banking facilities whereby cheque, among other forms of money, consolidated the modern monetary system including the granting of large credit.[39]

To enable the monetization of the colony meet its set objective, the British government created the West African Currency Board (WACB) in 1912 with its headquarters in London. The Board was to perform many functions, notably those of controlling currency supply, supervising over the currency interests of the colonial territories, and making special arrangements for minting coins authorised for circulation in the colonies. As part of the overall objective of the monetization agenda, the Board also created the Bank of British West Africa (BBWA) in 1896. The BBWA began operation in Nigeria in 1900. It was the forerunner of Standard Bank of Nigeria, now First Bank of Nigeria PLC.[40]

Once the task of monetisation was completed, other processes of capitalist penetration of the Delta area in particular and Nigeria became much easier. Let us now examine how monetization and other policies of the state facilitate trade.

4.3.3 Trade

Though trade in the pre-colonial economy of the Niger Delta was a major economic activity, the absence of a widely accepted legal tender hindered its expansion. With the imposition of the British currency on the Delta region, its economy like that of the rest of the colony, became intricately tied to the modes of capitalist accumulation in the metropolitan Britain. So, too, was the lopsidedness that characterized the trade relationship between the Delta area and Britain. A relationship in which the Delta area was assigned the role of producing palm oil/kernel, and crude oil, but net importer of British manufactured goods.[41]

Let us illustrate more on this with the case of palm products (oil and kernel) which was a major export of the Delta area to the metropolitan factories in Britain. The boom in the palm oil trade stemmed from its use for the production of margarine, pomade, candle among others. As a major exporter of palm produce, the Eastern Niger Delta for instance, accounted for 87 percent of the total national output of 190,623 tons of palm oil exported in 1955.[42] Yet, all that the area, like other palm producing areas in the western Nigeria, got in return in the form of imports, were candles, soap and margarine, all of which were produced from palm oil. Trade, conducted in this way, only turned the colonial economy of the Niger Delta into a consumerist one. Herein lies the imperialism of international trade. In fact, the structural dependence of the Delta region, which the pattern of trade had engendered was reinforced by the influx of British trading companies into the area.[43]

4.3.4 The Establishment of Monopolies

The 'scramble and partition of Africa' was one of the contradictions of European imperialism in Africa. As an intense struggle between and among international finance capitals from Western European countries, notably the United Kingdom and France, over the ownership and control of raw materials and markets in African colonies, it demonstrated how foreign capital pillage the economy of the continent. In Nigeria, which was exclusively dominated by British trading companies, its colonial economy was forced to produce what it did not consume, and vice versa. The United African Company and its subsidiary companies like John Holt and Patterson Zochonis particularly played a prominent role in the disarticulation of its economy[44]. Let us expantiate this with how the operations of the British trading companies impacted negatively on the economy of the Niger Delta area.

As trading monopolies, the basic concern of UAC, and its subsidiaries was to buy export produce from the farmers, which was mainly palm oil/kernel. In essence, they had no plan to invest in the growth of palm tress since the output from the peasant farmers met their demand. The problem was not really with not owning palm plantations as the policies of colonial capitalism on investment. For, foreign capital was not interested in investing in the colony, but extract raw materials. In addition to the monopoly of the trading companies over export crops, they also controlled what came into the colony as imports, and fixed their prices as well. Using the marketing boards as a go-between, the foreign companies exploited the peasant farmers by paying less than the price fixed for the export crops in the global market. All was to undermine the well-being of the peasant farmers who got far less proceeds if they sold directly to the companies.[44] An illustration with the basic export crops in the Niger Delta might be helpful.

As a major raw material input for the manufacturing of soap, cooking oil, feeds for cows, lubricants among others, palm oil/kernel were in high demand in western Europe particularly in the United Kingdom. For instance, with the invention of the locomotive engines and motor cars in the UK, there was great demand for rubber since it was used for the production of bushings, waterproofs and tyres among other parts. To obtain an adequate supply of these raw materials, the economic policy of colonial capitalism compelled the Niger Delta to specialize in producing palm products. Rubber tree was not only introduced into the area by the foreign companies, but they forced the people to cultivate it purely as an economic tree. In fact, because palm produce and rubber crumbs were got from palm trees and rubber trees respectively, they became economic trees, with penalty awaiting anyone found cutting them. Of the total national output of 607,000 tons of palm oil and kernel exported, and with a total value of £34.4 million in 1959, more than 75 per cent were produced in the Niger Delta with the remainder coming from the Western region of Nigeria. And of the 40,000 tons of rubber exported in the same year with value of £7.02 million, about 75 per cent also came from the Niger Delta.[45]

Furthermore, the mode of surplus extraction of the monopolies left the economy of the Niger Delta both export and import dependent with greater negative consequences for its capacity to develop along the lines of the aspiration of its people. As Jones rightly noted, geography of the Delta area made its trade imperative for the survival of its economy[46]. In the pre-capitalist economy of the Delta therefore, trade was prosecuted largely by the indigenes most of whom were middlemen for the foreign trading companies. It was from this group of local collaborators that the distributors of imported goods emerged in the colonial period.

But beyond acting as go-between for the foreign companies in the colonial period, the local business group had a weak capital base, which explained why it could not get really involved in production. It was all the more so because the foreign companies systematically denied them access to bank credit. Given the large size of the foreign trading companies, and their advantages of economy of scale, vertically integrated mode of production, and access to banks credits, it was easy to displace the indigenous business group. As foreign capital took control of the economy of the Delta area, the growth of the local business group was shunted, and local capital formation further declined. For instance, as distributors, they had their supplies controlled by the big monopolies, or their operation stifled with powerful mergers as means of checking them. A case in point was the formation of the Association of West African Merchants (AWAM) in 1937 comprising John Holt, UAC, PZ and UTC, with the sole purpose of dominating the buying of exports in the Niger Delta and other parts of the country.[47] That is not all as in 1947, the colonial state prohibited local from purchasing and marketing agricultural products.[48] These policies and actions of the colonial state place the economic prospects of the Niger Delta in the hands of foreign capital.

One of the hallmarks of the activities of the monopolies was the distribution of the cottage industries in Niger Delta. Cottage industries were dominant in the economy of the Delta area. They included native salt factories, cloth weaving, black-smithing, building canoes, local ginneries and soap factory. The colonial state policy made all of them illegal ostensibly to create market locally for foreign capital. As a major economic activity among the Ijaws, native salt was produced through a very simple process. Since the swampy mangrove forest which occupy a larger vegetational belt of the Niger Delta was salty, roots of the mahogany trees are cut, then dried, and later burnt. Thereafter, the ashes are collected, mixed with water, filtered and the filtrate boiled. The end product is an edible salt which is consumed by people of the Delta region.[49] As an occupation, it also provided employment for families and a regular source of income for those who were engaged in it. But all this was prohibited by the colonial state in order for John Holt, GBO, PZ, UAC among others, to market the edible salt imported from Europe. The consumption of the latter was discouraged on flimsy reason that it was unhygenically prepared whereas the real intention was for the protection of the local market against any competition. To ensure compliance, the colonial state mobilized its tools of oppression notably the police and staff of the native authority governments to invade the salt-producing communities across the Niger Delta, had their local equipments destroyed, and imprisoned offenders who still engaged in its production or sold it.[50]

The local ginneries were not spared also. Like the native salt factories, they were located in the coastal areas where the wild raffia palm trees from which the wine used for the distillation of the local gin, were grown in the area. The local gin popularly called 'ogogoro' in the Niger Delta was seen by the trading foreign companies as capable of competing favourably with their imported spirits like the Gordon's Dry Gin from Europe. This was all the more so as investigation had shown that the majority of the people of the Delta area and across Nigeria generally, preferred to drink the local gin, 'ogogoro'. To secure the local market for imported gin and other alcoholic drinks, the production of 'ogogoro' and its sale were prohibited by the colonial state, prescribing prison terms ranging from six months to three years for anyone caught making, marketing and consuming it.[51] The local soap factories were equally shut down by the colonial state apparently to create for imported soap like palm olive, sunlight and lux which were brought into the country by the trading companies.[52]

Not only did the annihilation of pre-colonial cottage industries in the Niger Delta by the colonial state furthered the integration of its economy into the British capitalist fold, labour, hitherto employed in these economic activities was displaced and forced by the dictates of capital to sell itself at whatever price in order to survive. This was to constitute the origin of the pool of urban cheap labour in the Niger Delta from which Shell recruited its casual workers when it began oil exploration there in 1937. It was from the same pool of cheap labour that the UAC got its casual labourers when it slightly shifted business from mainly buying agricultural produce into the production of these crops to complement its export of raw materials to the metropolitan factories. After the end of World War II in 1945, there was sudden rise in the demand for agricultural produce in Europe. Suffice it to note that, it should be construed to mean UAC venture into production to develop the colony. Rather, the United African Company had veered into the growth of rubber and palm trees in Oghara near Sapele in the present day Delta State not with the intent to invest and improve the economy of the area as the operation of the plantation was still disarticulated from the local economy. If anything, complemented the declining output of the peasants in the face of the rise in demand. So, too, was the intention of the African Timber and Plywood Company (AT&P), - a subsidiary of the UAC, based in Sapele, when it began to semi-process its major raw materials (timber) into sewn plywood and making of flush doors among others. All was aimed at increasing the quantity and quality of the timber products being exported, but not to help the local economy.[53] All this means that the multiplier effects of the operations of the big trading companies were reaped in their home factories as the semi-processing of timber by AT&P

for instance, was not aimed at industrializing the economy of the Delta since there were no ancillary industries set up for that purpose either by the state or the company in its host economy. This trend was continued with when Shell began operation there. Let us now examine how taxation aided the process of exploitation of the Niger Delta by foreign capital.

4.3.5 Taxation

Like all its colonies in Africa and elsewhere, the British had no interest in the economic development of Nigeria including its Niger Delta. Perhaps, this explained why it failed to provide funds to administer the colony. Rather, taxation was seen as by the British as a major source for raising money, for the running of the colony. How was it carried out in the Niger Delta, and what was its impact on the economy of the area?

Hopeful of the success of taxation in Northern Nigeria where the feudal system eased the collection of taxes by the emirs, it was introduced into the Niger Delta with similar optimism in 1927[54]. Such optimism was partly rooted in the hope that with the monetization of the colonial economy of Nigeria inclusive of its Niger Delta, collection of tax was made a lot easier. It was however, met with stiff opposition for reasons not too far to seek. First, was the relative absence of feudalism in the Niger Delta. With the majority of the nations across the Delta region republican in nature, there people were not really imbued with the culture of paying tribute to kings, the exact opposite of which existed in the monarchies. Second, and perhaps, more important, was the rumour that women were to be taxed. A rumour that led to the spread of anti-tax riots in the communities around Warri in 1927, and two years later, at Aba, all with severe casualties.[55] The warrant chiefs were particularly entrusted with the task of enlisting of eligible adults for the payment of income tax with its collection left with the officials of the native authority. It was a duty that set the people in collision course with the chiefs who were seen as aiding the colonial government to exploit the people. Resistance by the people was however, repelled with the speedy intervention of the colonial police. In fact, the anti-tax riot of 1927 in Warri and that of 1929 at Aba, were speedily quelled by the colonial police.[56] In spite of the anti-tax riots, income tax was still imposed on the people's proceeds from which accounted largely for the revenue of the colonial state. Indirect tax, which came mostly in the form of import duties, and licences, was another main revenue source of the state.[57] The issue at stake is not so much with taxation as a modality for the underdevelopment of the Niger Delta area.

Like other modalities of exploitation, taxation helped consolidate colonial capitalism in the Niger Delta. In playing this role, overthrow of the pre-capitalist mode of production in the Delta, taxation was used as a major tool for enforcing wage labour and creating pool of urban cheep labour. Since income tax was levied per head the people really had no real option than to work in order to earn enough money a large portion of which was used to pay tax. As McPhee rightly noted,

> At first, Africans had to be press ganged into the (Public Works) Service. Later, they became eager workers when they discovered that instead of labouring at their own charges, they were given silver discs which could fetch them all manners of desirable commodities at the nearest 'store' ... Direct taxation, conversely, compelled Africans to accumulate a certain amount of the same silver to avoid the displeasure of the government.[58]

McPhee also aptly captured some of the negative impact of taxation on the colonial economy of the Niger Delta in particular, and Nigeria generally. That is, 'to avoid the displeasure of the government' which meant paying income tax in order not to be imprisoned by the colonial state. In search of money, especially to pay tax, some people who could not farm, were force to migrate to Warri, Sapele, Port Harcourt among other cities in the Niger Delta looking for daily paid casual job. Yet, others who could not move to the cities, migrated seasonally from one community to another particularly during the planting season to work on the land, thereafter, sold their harvest the proceeds from which were used to pay tax and meet other social needs. This type of migrant labourers were most common in the hinterland areas of the Niger Delta, notably among the Urhobo and Isoko farmers in its western part. For those who refused to migrate, but unable to pay income tax, they usually took to the bush on the sight of the colonial tax collectors. It is pertinent to add that, although the ultimate aim of subsistence production in the colonial period of the Niger Delta and Nigeria generally, was to earn money to pay income tax, part of the wages was used to meet other basic human needs like clothing and housing.

A more detailed discussion on the colonial taxation policy, its effect on labour, and how Shell took full of them are taken up in the course of the discussion. It is in order to examine how colonial policy on transportation further structured the economy of the Niger Delta for capitalist exploitation.

4.3.6 Transport

As noted earlier, one of the unique geographical features of the Niger Delta is the network of rivers that criss-cross its terrain. Rivers that ease trade within the area, and between the region and the rest of the world. In fact, the relative success of slave trade in Nigeria was due in part to the Niger Delta where the rivers and creeks made the movement of slaves from the hinterlands to the coastal areas and then overseas, very easy. Significant towns like Bonny and Old Calabar in the pre-colonial history of the Niger Delta owed their importance to being coastal cities where slaves were bought and sold.[59]

With the imposition of British rule, these natural means of transportation were not particularly disrupted. Rather, the foreign trading companies still used them to facilitate the export-import trade. They did not use all the existing rivers and creeks which made slave trade to blossom in the area. Instead, few coastal towns were selected and developed into ports for the loading and off-loading of both exports and imports. The choice of which coastal town to develop or abandon, was guided by the calculation of the foreign companies; calculation in which little or no attention was paid to the development of the host communities. Suffice it to note that, transportation routes were developed to further the capitalist penetration and exploitation of the Niger Delta. While cities like Port Harcourt, with its deep natural habour was further developed to facilitate the export and import business, hitherto famous coastal town like Bonny which was noted for slave market, was abandoned. Other important 18th Century coastal towns like Akasa, Brass and Opobo in the eastern part of the Niger Delta were equally forgotten.[60] In the western part of the Niger Delta cities like Warri, Sapele, Forcados, and Koko, were developed to the neglect of Ebrohimi, former commercial nerve centre of Chief Nana located along the Benin River.[61]

The implications of the colonial policy on transportation for the economy of the Delta area were mixed. The new port-cities developed by the colonial state and foreign capital had basic social infrastructures like pipe-borne water, electricity and hospitals all of which furthered capitalist exploitation of the area, but not to bring about the increase in the material well-being of the majority of people who live there. As ports created by the big foreign companies and the state, they had collection centres and depots warehouses for storing both exports and imports. All was to turn them into commercial nerve centres in the Niger Delta to which migrant unskilled labour was attracted in search of wages.[62] Most of the coastal cities like Aravwarien and Okpare which served as depots were abandoned by the UAC as soon as there was decline in the palm oil trade just as Ebrohimi was forgotten with the abolition of slave trade. Furthermore, whatever could be regarded as benefits from the colonial transportation policy eventually led to the creation of poles of growth and stagnation across the Delta area; a situation akin to the 'centre-periphery thesis' of Gunder Frank. Within Warri, for instance, this was evidenced in the urban slums around Agbasa and Effurun where the urban proletariat and other members of the down-trodden class sojourned representing the pole of stagnation, while its reserved areas a resemblance of what Segun Osoba called the Government Reserved Area (GRA), where the officials of the colonial state and the trading companies stayed.[63] The colonial mode of transportation also heightened the rural-urban pull, which left the bulk of the economic activities in the rural areas in the hands of the old. Shell cashed in on all this when it began operation there. It continued with the trend by constructing its access roads to further the penetration of the Niger Delta for exploitation.

4.3.7 Education

The colonial state had introduced education mainly to exploit foreign capital needed some level of skilled labour for it to exploit the colony. Raising manpower locally therefore, was aimed at reducing the importation of expensive expatriate labour from Europe. Thus, the colonial state established institutions where technicians, plumbers, medical auxiliaries, clerks, book-keepers and interpreters among other semi-skilled personnel, were raised to service the economy.[64]

In the Niger Delta for instance, both the colonial state and Christian Missions established most of the primary and secondary schools. As a mode of exploitation, the curricula of the schools were on moral codes, obedience, history of great European explorers and warriors and loyalty to government. Teachings in the Mission schools also preached against dishonesty, theft, protests to constituted authority and emphasis on the fear of God. All these were meant primarily to ensure that their graduates accepted the colonial state and its exploitative policies without contest. Notable Mission Schools in the Niger Delta area were the Hope Waddel Training Institute, Calabar which was established in 1895 by the Church Missionary Society (CMS); St. Thomas Teacher Training College, Warri by the Roman Catholic Mission; and James Welch College, Emevor by the Anglican Mission.

As colonial capitalism took firmer root, greater skilled labour was needed as graduates of these missionary schools could not really meet the manpower needs of foreign capital. As a result, Yaba Higher College was established in 1920 ostensibly to train medium-level technical staff needed to facilitate the exploitation of the economy of Niger Delta. By 1948, the University of London established a College at Ibadan. But this time, it was meant for higher manpower needs of the colonial state and foreign capital. To further entrench the legacy of exploitation in the colony, it was largely those Nigerian who served in the colonial government admission into the college for their children. The fact that they were relatively wealthier made it all the more so. Little wonder the wards of warrant chiefs and traditional titled holders in the Niger Delta were largely those who had university education. Notable ones in the western Niger Delta included: the Prests and the Rewanes, the Don Domingos. The situation in the eastern Niger Delta was not significantly different explaining why the offsprings of King Pepple of Bonny was among the early ones who received western education.[64] Some of the nationalists like Chief Mowoe, was able to train his children to the university level.

As noted earlier, the same group of educated elites raised by the colonial state to ease the capitalist exploitation of the Niger Delta, and Nigeria generally, that turned out to be its grave-digger. The move to sack the colonial state was informed by several important reasons. First, the local political class, having understood the ideology and policy of the colonial state, began to clamour for self rule since it was unwilling to share political power, talk less of transferring it. Second, colonial capitalism created a pool of urban labour which always rebelled against capital at the slightest chance. Conscious of its exploitation, labour began to show its dissatisfaction with colonial capitalism. The need arose therefore, for both of them, (the political class and labour) to collectively overthrew the colonial state. Third, was the discrimination between colonial capitalists and the indigenous entrepreneurs. Foreign banks never granted loans to the local businessmen in the Niger Delta for lack of collateral. Land, which ought to have been one important collateral security for loan, was rejected by the banks because of its communal ownership. Yet, foreign companies obtained bank loans largely by goodwill. All was to constitute a major contradiction of the colonial state which the local political class mobilized in its canvass for self rule. Notable among them in the Niger Delta, were Chief Yamu Nuna, Mr. J.G. Ako, Chief Arthur Prest and Chief Festus Okotie-Eboh most of whom had gone into full scale politics by the early 1950s.. Chief Festus Okotie-Eboh for instance, was to become the first Federal Minister of Finance in 1960, the year the colonial state was sacked in Nigeria.

However, unanimous as the nationalists seemed in their struggle against colonialism, as a class, the power base was ethnic and national. As it turned out, they were not out for a change but to inherit the colonial policy and structures of exploitation. In essence, together with foreign capital, those entrusted with the Nigerian state furthered the exploitation of the Niger Delta, particularly when Shell began crude oil exploitation there.

4.4 Conclusion

The nature of capitalist development that the forces of colonial capitalism had promoted in the Niger Delta before the advent of Shell, was such that tied it in a dependent relationship with Western imperialism. However, though the origin of the pillage of the Niger Delta predates Shell, the relative reckless manner in which the company had explored and produced crude oil there wrecked more havoc on the environment and economy of its immediate host communities than when it was a producer of palm oil and kernel. The fact that the production of crude oil amounted to the physical destruction of the balance of ecosystem particularly its environment, naturally follows that its consequences were worse. Therefore, the rest part of this study is devoted to an analysis of Shell's mode of operation and its socio-economic and environmental impacts on the Niger Delta. In doing this, the next chapter first examines the activities of Shell as oil capital, in the context of the global petroleum industry in order to deepen our understanding of why the company shifted business to Nigeria and the nature of its operation in Niger Delta.

Notes and References

1. There is no doubt that the British government used the Royal Niger Company to accomplish its economic adventures in the Niger Delta area. The activities of the company cannot be equated with the formal colonization of Nigeria and its Delta area by the British. For details on the activities of the RNC, see J. Flint, *Sir George Goldie and the Making of Nigeria*, for the operations of the RNC in the Delta area in particular, see Obaro Ikime's *Rivalry in the Niger Delta ... op.cit.*, his *The British Conquest of Nigeria ... op.cit* and Chapters 6 and 7 of Dike's *Trade and Politics in the Niger Delta ... op.cit*
2. The cowries, salt and some other commodity - currencies were still circulated side by side with the gold and silver discs until 1880 when

Ordinance No.2 was passed to demonetize certain currencies. The introduction of paper money was still rejected by the people of the Delta area after the Ordinance had come into force because they had no confidence in the British notes and could not be equated with the worth of their goods. See the *Niger Coast Protectorate Blue Book*, 1897-8, CO 462/2; A.G. Hopkins, 'The Creation of a Colonial Monetary System: The Origins of West African Currency Board, '*African Historical Studies*' Vol.3, 1, (1970); For a more refreshing account of the economic history of Nigeria, see Olufemi Ekundare's *Economic History of Nigeria 1860-1960*, (London, Methuen and Coy, 1973). The first bank in Nigeria which began operation in 1891 the British Bank of West Africa (an affiliate of the Standard Bank of England) with its main office in Accra, Ghana. Proper banking activities did not however start until the early 20th Century when the Delta area had been fully re-integrated into the British financial system and by which time the use of British notes and coins as legal tender for transaction and payment of taxes among others had been firmly rooted. See also, Olufemi Ekundare's *Economic History of Nigeria ... Ibid.*

3. If the 'Niger Delta opened the gateway to British Imperialism in Nigeria,' then it was the Royal Niger Company which was entrusted with both political and economic power that made it possible. For one thing, all the heavy military attacks that the Delta area faced, happened during the reign of the RNC. See for details, Ikime's *British Conquest of Nigeria op.cit.*; Ikime's *Rivalry in the Niger Delta op.cit.*; and J. Flint, *op.cit.*
4. There is still dearth of written history on the pre-colonial period of the Niger Delta. In fact, the few available materials on the Delta at this period were documented by European adventurers who were not strictly trained enthnographers and anthropologists. Even at the time Kingsley Dike wrote his book titled *Trade and Politics in the Niger Delta*, he faced the same problem and acknowledged it as well. To that extent, the researcher relied heavily on field work during which he had oral discussion with selected respondents who are mostly elders and retired colonial officers.
5. Eme Ekekwa made a clarification of this in his *Class and State in Nigeria*, (London, Longmans, 1986).
6. See Dike, *Trade and Politics ... op.cit.*
7 See Alogoa, *The Small Brave City ... op.cit.*; Ikime, *Niger Delta Rivalry ... op.cit.*; Dike *op.cit.*; and Jones, *The Trading States of the Oil Rivers ... op.cit.* Although colonialism brought artificiality into the traditional political structure of the Niger Delta, the tradition and customs of the people have refused to die.

8. For more details on the political system of the Urhobos, see Obaro Ikime's *Niger Delta Rivalry ... op.cit.*
9. The 'elotus' was (and is still the age group that provides security for the Urhobo community and reservers the right to declare war on any other community.
10. In the monarchies, there were usually houses assigned with a particular role. Thus, there was the 'War House", the 'Canoe house' among others. For details, see Jones, *The Trading States of the Oil Rivers ... op.cit.*
11. See, in particular, Ikime's *Niger Delta ... op.cit.*; also based on the discussion with some elderly people in the Urhobo speaking communities.
12. There is no doubt that the Delta area had metamorphosed from one mode of production to another. For instance, from communal to slavery, until it finally gets to the capitalist mode of production. This Chapter is concerned with the inception of the capitalist mode or production. For details, see '*The Origin of the Family, Private Property and the State*, (Moscow, Progress 1974) Marx's *Capital op.cit.*; Lenin's *The Development of Capitalism in Russia op.cit.* See also, Walter Rodney's *How Europe Underdeveloped Africa*, op.cit.
13. For detail on the capitalist mode of production, see Marx, *op.cit.*; Claude Ake's *Political Economy of Africa ... op.cit.*
14. Based on the author's interviews with the Ijaws and Urhobos, among others ethnic groups in the Delta. See also, G.T. Basden, *Among the Ibos of Nigeria in 1921* cited in Eme Ekewe's *Class and State in Nigeria op.cit.*; Obaro Ikime, *op.cit.*, Joseph Alogoa, *op.cit.*
15. For details on the land ownership among the Ibo speaking areas of the Eastern Delta, see Badsen, *ibid*; See Ikime, *ibid*; For the situation among the Urhobos, see also, Frederick Lugard, *Political Memoranda: Revision of Instructions to Political Officers on Subjects Chiefly Political and Administratives* cited in A.H.M. Kirk-Green ed., *Crisis and Conflict in Nigeria: A Documentary Source Book 1966-1969*, (London, Oxford Press, 1971).
16. For details on the Theory of Surplus Value, See Marx *op.cit.*
17. *ibid*; See also, Ake's *Political Economy ... op.cit.*
18. See Dike, *Trade and Politics ... op.cit.*
19. Obaro Ikime's *Rivalry in the Niger Delta* is a good source though, the researcher still had discussion with selected members of the Urhobo-speaking areas of the Delta.
20. For details on the slave-holding mode of production, See Engels' *The Origin of the Family... op.cit...*; and Lenin's lecture *The State.*

21. See Obaro Ikime's *The Merchant Prince of Niger Delta,* (London/ Ibadan: Heinemman, 1968).
22. *ibid,* and also Dike's *Trade and Politics... op.cit.*
23. *ibid.*
24. See G. Jones, *The Trading State of the Oil Rivers... op.cit.*
25. See Dike *op.cit*; Ikime *op.cit.*; and Jones, *ibid.*
26. See F. Engels, *The Origin of the Family... op.cit.*
27. *ibid*
28. The entity that become Nigeria today was the result of the amalgamation of the Northern and Southern Protectorates by Frederick Lugard in 1914. See Lord Lugard, *Political Memoranda,* pp.296-318; and A.H.M. Kirk-Green, *Lugard and the Amalgamation of Nigeria: A Documentary Record.* (London: Frank Cass, 1968) pp.20-26.
29. See R.O. Ekundare, *An Economic History of Nigeria 1860-1960, op.cit,* p.12.
30. See Toyin Falola ed., *Britain and Nigeria: Exploitation or Dependence,* (London: Zed Press, 1987).
31. See, for instance, Eme Awa, *Federal Government in Nigeria* (Berkley: University of California Press, 1964), p.46. The analysis of Awa was applicable to the Delta area.
32. See for example Adebayo Adedeji, *Nigerian Federal Finance: Its Development, problems, Prospects,* (New York: Hutchison, 1969).
33. See CO/520/128 Confidential Enclosure of 29 December, 1913 by Lord Lugard to the Secretary of State for the Colonies, 11 December, 1913. The Mineral Ordinance was enacted in 1912 the amended in 1913. For details, see CO/520/128 Confidential 14462 of 29 December, 1913.
34. *ibid*; See J. Iloeje, *Geography of Nigeria ...*
35. Obaro Ikime treated the role of the Royal Niger Company in his *The Fall of Nigeria ... op.cit.*
36. See, for instance, Michael Crowder and Obare Ikime eds., *West African Chiefs: Their Changing Status Under Colonial Rule and Independence,* (Ile-Ife: University of Ife Press, 1970). Throughout the Delta area, there were stories told of chiefs who levied fines ranging from 200-400 tuber of yams and court records showed half of them was consumed by the warrant chiefs. See Raymond Clough, *Oil Rivers Traders: Memoirs of Iboland,* (London: c. Hurst, 1972), pp.38 and 84.
37. *ibid.*
38. See The Importation of Cowries Proclamation No.6 of 1904.

39. See Olufemi Ekundare's *Economic History of Nigeria ... op.cit.* Before 1900, the commodity-currencies in the Delta were already depreciated by the RNC. The iron bar for instance, depreciated steadily throughout the 19th Century. For details see, the *Niger Coast Protectorate Blue Book 189708*, CO462/2.
40. *ibid*, Ekundare
41. See Dike's *Trade and Politics in the Niger Delta ... op.cit.*; and also *Annual Abstract of Statistics*, Lagos, FOS, 1964, p.33.
42. See Chapter Five in Ekundare's *Economic History of Nigeria ... op.cit.* Claude Ake's *A Political Economy of Africa, op.cit*; and Chapter Four of Toyin Falola's *Britain and Nigeria op.cit. op.cit.*
43. See Kilby *op.cit*; Ake (1978) *op.cit* and *Ekundare, op.cit.*
44. *ibid.*
45. See, *The Economic Survey of Nigeria, 1959* (Lagos, Government Printers, 1959), p.27.
46. See G.I. Jones, *op.cit.*
47. See C. Beer *The Politics of Peasant Groups in Western Nigeria 1969* cited in Ekekwe *op.cit.*
48. *ibid.*
49. Based on author's interviews Ijaw people in Burutu, Patani and Warri in August 1989.
50. *ibid* (Although its market was limited, the people still produce and consume native salt in the Delta area).
51. Based on field trips to the 'ogogoro' producing areas of Bayelsa, Delta and Rivers States in 1990 and 1997.
52. Based on author's field trips in June 1989 and August 1991 to Arawvarien, a notable traditional soap making community in Ugehelli council area.
53. Based on author's interviews with some officials of AT&P Sapele in August 1989.
54. See for instance, Obaro Ikime, 'The Anti-Tax Riots in Warri province, 1927-28 *Journal of the Historical of Nigeria (JHSN) III*, 3, (June 1967).
55. *ibid* for details on the Warri Anti-Tax Riots; see Harry Gailey's *The Road to Aba...* (New York: New York University Press, 1970).
56. See A. Afigbo's *The Warrant Chiefs*, (London, 1972). p.6.
57. Ikime, *op.cit*; and Gailey, *op.cit.*
58. See McPhee's *The Economic Revolution in British West Africa*, p.238 as cited in Ekekwe *op.cit.*

59. For details on the trade in the Delta and the role of water transport particularly during the period of slavery, See for instance, Dike *op.cit.*
60. See Ikime's *Merchant Prince of the Niger Delta ... op.cit.*
61. See, for instance, Akin Mabogunje's *Urbanization in Nigeria*, (*London*: University Press, 1968).
62. *ibid.*
63. See Chapters One, Two, and Three of this work for details on the center-periphery thesis.
64. See Babs Fafunwa, *History of Education in Nigeria* (London: Allen Unwin, 1974). Lord Lugard, *Political Memoranda op.cit.* p.27; and Otonti Nduka, *Western Education and Nigerian Cultural Background* (Ibadan: OUP, 1964).
65. Based on author's interviews with numerous early educationists and officials of colonial government in the Delta area in 1988, 1989, 1990.

Chapter 5

Shell Petroleum Development Company and Capitalist Development in the Niger Delta

5.1 Introduction

This chapter critically examines the political economy of the operations of Shell in the Niger Delta. Before doing so, there is need to first understand the nature of the global capitalist petroleum industry taking cognizance of the contradictions that have characterized the sector since 1945 part of which explained why Shell, second largest oil company in the world after Exxon, shifted the business of oil production to the Delta area of Nigeria among other countries in the Third World.

5.2 Antecedents of the 'Oil Majors' in the Global Petroleum Industry

The early origin of the world petroleum industry can be traced to the American and Russian oil companies. By 1850, some private investors of American origin had begun the exploration and production of crude oil in the Southern part of the country. A decade later, that is, by 1860, their Russian counterparts joined the business of crude oil exploration and production. During this period, the imperial powers of Europe namely Britain, France, Spain, Italy did not show any significant interest in the oil industry as coal was a dominant energy source. Even then, the production of crude oil was still limited to the satisfaction of local energy needs.[1]

By the turn of the 20th century, crude oil became the most important source of energy in the world for reasons not too far to see. First, it had successfully powered the British war machines during the World War I, a war in which it enabled the British to win: and thereafter, it finally replaced coal as about the cheapest energy source with no known substitute since then. Second, is its demand in the modern industrial age which emphasized the constant need for energy so that capital can reproduce itself profitably. It was more of the latter reason than the former, that compelled the individual investors, private and public corporate capitals and governments to invest in the petroleum industry.[2] By 1910, these investors had incorporated various oil companies in the United States of America, and Western Europe. In the United States of America, for instance, the Rockefeller Trust which was founded by John D. Rockefeller, had set up the Standard Oil Company with monopoly control over the supply and price of petroleum products. In Britain, Marcus Samuel, had, within the period also, incorporated the Shell Trading and Chemical Company which was dealing on chemicals, petroleum products sea transportation among others.0000 Henri Deterding, a Dutch, floated the Royal Dutch Oil Company whose major business was the sale of petroleum products. These were the major early oil companies that later expanded their operations to become the forerunners of the dominant oil monopolies like Exxon, Shell which have been controlling the world petroleum industry since 1945.[3]

Between 1860 and 1910, these private oil companies had also expanded their scope of operations, which had spread to most parts of the world. Because of their integrated network of operations too, they were able to control both the supply and price of petroleum products. For instance, on account of the size and monopoly position of the Standard Oil Company, it was able to control the oil market in the United States of America. A dominant position which became so clear to the American government that, it would be risky allowing the oil business of the country, and by implication, the prospect of its economy incessant fuel shortages which not only characterized the American economy, but also, about wrecked it. As a consequence, the American government enacted the Anti-Trust Law in 1911, prohibiting the formation of oil cartels in the country.[4] In that same year (1911), the US Supreme Court ruled that the Standard Oil Company should be dissolved and that all its subsidiary companies to operate as independent companies.[5]

As a result of the 1911 Anti-Trust Law, the Standard Oil Company New York, one of the companies under the Rockefeller Trust, merged with the Vacuum Oil Company, and was known first, as Socony-Mobil, and later renamed, as Mobil Oil Company. The Standard Oil Company of California (Socony) was another Rockefeller's Trust companies which also became independent. Exxon, (formerly Esso), which is today's largest oil company in the world, was also one of the companies under the Rockefeller's Trust. In fact, Mobil, Standard Oil of California, and Exxon were all offsprings of the then Rockefeller Trust. But Gulf and Texaco oil companies also had American origins, but were formed outside of the Rockefeller oil empire. Gulf Oil Corporation (Gulf) for instance, was started from the capital raised by some businessmen and bankers in Pittsburgh and Pennsylvania to finance an oil venture which led to the discovery of oil at Spindletop in Texas state. Texaco, on the other hand, which was originally an oil company based on Texas. developed in its own way.[6] Out of the seven major oil companies that have to dominate the world petroleum industry, which are referred to here as the 'oil majors', five of them were of American origin. These are Exxon, Mobil Oil Corporation, Standard Oil of California, Gulf Oil Corporation, and Texaco Incorporated.

Although the Anti-Trust Law of 1911 in the USA led in part, to the incorporation of British Petroleum (B.P.) by the British government, the birth of the company was made imperative by the outbreak of the World War in 1914; a war in which the use of petroleum fuel to fire the British naval ships facilitated its victory.[7] One of the effects of the Act on the petroleum industry in Europe was the erosion of the monopoly which Shell International Trading and Transportation Company and the Royal Dutch Oil Company had over the production and sale of petroleum products in the United Kingdom. With the entry of BP into the petroleum industry in the U.K. and other parts of Europe, investment capital in the industry began to get overconcentrated, while the hydrocarbon reserve decreased. In response to the crisis which monopoly capitalism had engendered in the industry, the two companies merged their interests while retaining their separate identities to form the Royal Dutch/Shell Group of Companies in the ratio of 60:40. It was this merger that then moved their various oil business to the countries of the Third World.[8] One of the basic advantages of the merger is that, it enabled them to pool more resources together particularly as they shifted oil business to the Third World, where they expected to enjoy a higher returns on their investments.

By the 1920s, seven major oil companies had dominated the world petroleum industry namely, Exxon, Mobil Standard Oil, Gulf, Texaco, Royal Dutch/Shell Group (Shell for short), and British Petroleum. Together, they operate across the world with the exception of most of the centrally planned economies, controlling over 80 per cent of all oil productions, 70 per cent of the world refinery capacity, and 50 per cent of the oil tankers.[9]

Outside the seven major oil companies, are other independent oil companies otherwise known as the Independents that are directly engaged in the various stages of the petroleum industry. Because of their limited financial resources which, in turn, influenced their size of operations, they hardly hold sway in the global petroleum industry. Some of them include the French Total Group and the Italian ENI.[10] However, the purpose of the anti-trust laws and other government's interventions to rid the global petroleum industry of monopoly capitalism, did very little to really control them. They led to the restructuring of the petroleum industry and the oil companies in their home countries. The ownership structure of the oil companies which in the post-1911 period was, rather than woven around an individual investor like John Rockefeller, became corporate bodies with many independent investors as shareholders. For instance, the Standard Oil Trust which used to be one of Rockerfeller's financial empires, had its equity share of Rockefeller himself, reduced to a minority position as more shareholders bought more shares in the company. The same was true of Shell International Trading and Transportation Company and the Royal Dutch Oil Company where their shareholders included individuals, banks, churches and insurance companies. All has reduced the shares of their pioneers, Marcus Samuel for Shell, and Henri Deterding Royal Dutch Oil Company) to the barest minimum. But that is about that was achieved as the seven major oil company still control oil production, determining the prices of its bye-products across the world petroleum market.[11]

There is no doubt that the recapitalization of the major oil companies as a result of the anti-trust law was to place more finance capital at their disposal. But the problem that soon emerged, was that surplus investment funds began to compete for the limited market outlets and raw materials in Europe and America. This led to the over concentration and centralization of investment funds in the oil industry, which, in turn, compelled the oil companies to move business to the countries in the Third World, led to the establishment of spheres of influence and sharing of the market outlets among themselves. It also enabled them to fix prices of petroleum products, reach general agreement on what should be the policy of the petroleum industry at the global level.

In shifting oil business to the Third World, the big American oil monopolies notably Exxon, Texaco Standard Oil and Socony concentrated in Mexico and Venezuela. In the Persian Gulf, the oil concessions were shared among the British and Dutch oil companies. The British government had, through the BP, backed up William D'Arcy, an investor, by providing him with financial assistance with which he established the Burmah Oil Company in the Persian Gulf. The British government also owned shares in the Anglo-Persian Oil Company. In fact, before the outbreak of the World War I in 1914, the British had united all European oil interests in the Persian Gulf to keep the Germans out of the oil business. This was to explain why the Germany's war machines and naval ships were short of oil and gas causing the defeat of Germany in that war.[12]

The world economic depression of the late 1920s which set in after the end of the World War I, left the economy of the United States of America particularly badly hit by another round of acute shortage of crude oil and its bye-products. This was due to the mass acceptance of automobiles and other energy demands of the world. Much more alarming, was the report of the US Geological Survey that alarmed the country that its hydrocarbon reserves were drying up very fast.[13] In response, large American oil monopolies, notably, Standard Oil and Socony, moved to the Persian Gulf where the Anglo-Dutch oil companies had all the probable oil fields in their hands, under their management and control, or financed by them.[14]

For the first time after the Anti-Trust Act 1911, the American government intervened in the world oil industry with a threat to cut off the New Jersey oil supply to Britain if American oil companies were not allowed into the Persian Gulf.[15] As a result of American diplomatic pressure and threat, on the British, the Red Line Agreement was signed in 1928 by the British, French, Dutch and American governments. The Red Line was drawn to show the pre-1914 Ottoman boundaries in the present-day Persian Gulf with the hope that it would settle once and for all, the Middle East oil concession agreements. With the agreement in force, British Petroleum, Royal Dutch/Shell, and the French CFP had a total of 71.25 per cent of the oil concessions, while, the American oil companies (Standard Oil and Socony) jointly had 23.75 per cent, and the independent Gulberkians took the remaining 5 per cent. By this ratio, the 'oil majors' still controlled over 90 percent of the oil concessions in the Gulf.[16] Even after the Italian corporatist philosophy towards the oil industry which led to the formation of Azienda Generate Italian Petroli (AGIP) in 1926, and the Spanish effort which culminated in the setting up of the state-controlled Compania Arrendataria del Monopolio de Petroleos (CAMPSA) in 1927, the dominance of the major oil companies in the world oil industry still remains till date.

With the outbreak of the World War II in 1939, the 'oil majors' once again, came under another threat by both the British and American governments. For instance, the Roosevelt administration had agreed with the British government to coordinate the oil industry in order to avoid the experience of the World War II during which 'oil majors' controlled oil production and the supply of fuel and gas to the allied forces. But the inter-governmental agreement between the USA and Britain was rejected by the oil majors on the ground that, considering the intricate and complex nature of oil production and supply, it would be uneconomic as well as risky, to allow soldiers and government into the business. In the alternative, they suggested that with a better liaison between them and the national bodies set up during the war for that purpose, the flow of oil would not be interrupted. Britain, because of her huge oil interests, supported the position of the 'oil majors', which were subsequently entrusted with the task of supplying oil to the Anglo-American war machines during World War II.[18]

After the end of the WWII in 1945, the Red Line Agreement was revoked. This became necessary as a result of the over-concentration of the oil companies (including the independents) and the resultant depletion of hydrocarbon reserves in the blocks under the Agreement. All this compelled the oil companies to continue the search for crude oil in areas like Saudi Arabia, Jordan, Bahrain that were outside of the Red Line. In 1946, Standard Oil, in conjunction with Socony, found their way into Saudi Arabia thereby revoking their membership of the International Petroleum Company (IPC), which theoretically barred them from seeking oil outside the boundary of the Red Line Agreement of 1928 in the Persian Gulf. Thus, the way to the oil land of the Middle East was opened to more oil companies, particularly oil companies from the United States of America[19].

In Africa, particularly in Nigeria, it was after the end of World War II that Shell Petroleum Development Company, a subsidiary of the Royal Dutch/Shell Group, actually began full scale oil exploration and production after an initial futile attempt in 1937.[20]

As a broad part of the 'scramble and partition of Africa',[21] the Mineral Act enacted by the British in 1914 as noted earlier, reserved the right to exploit mineral in Nigeria only for companies of British origin. This explained why Shell joined interests with BP in order to explore and produce crude oil in Nigeria, then a British colony. In spite of this pressures from American and other European oil companies on the colonial state for permission to prospect for crude oil in Nigeria, it was not until 1960 when Nigeria gained political independence that Mineral Act was amended to allow Euro-American oil companies into the country's petroleum industry. This issue is discussed in detail in the course of this chapter.

One hallmark of the post-1945 event in the petroleum industry was the anti-foreign oil companies feelings by the oil-producing Third World countries. The role of oil in the two world wars, the increasing demand for oil in the new industrial age, and return of oil experts from European and America to their home countries in the Third World fuelled the anti-oil companies attitude. That is not all as it was increasingly becoming clear to some governments of oil-producing countries in the Third World that, it was high time they had a decisive say in the petroleum industry rather than leaving the foreign oil companies to cell the shot. Oil-bearing countries like Libya, Algeria, first initiated the fight to break the monopoly of their petroleum industry by the 'oil majors'[22]. They were later joined by Egypt. Other Middle East oil-producing countries were not left out in the struggle to wrest control of the oil sector from foreign capital. In 1951, Mossadeq nationalized all foreign companies operating in Iran as part of his anti-oil companies campaign. Although three years later, his regime was overthrown and replaced with a conservative government that was pro-foreign oil companies, it served only to fuel the campaign against the 'oil majors' by other oil-producing nations in the Third World notably, the Arab oil states of Saudi Arabia, Kuwait, Quata, Yemen, Iran among others.[23] In Africa, Algeria and Libya took the lead as noted earlier.

By 1960, the anti-foreign oil campaign had crystalled into the Organization of Petroleum Exporting Countries (OPEC). So, what began as an anti-oil company campaign in the early 1950s had culminated in the formation of an oil producers organisation in 1960.[24] By 1971, Nigeria joined the Organisation. The formation of the OPEC marked a turning point in the history of the global petroleum industry because it sought to threaten the monopoly control of the industry by 'oil majors'. Whether the birth of the Organisation led to a significant control of the industry, especially the oil technology by its member-states, shall be discussed later in the light of Nigeria's experience with oil exploration and production.

So much discussion on the oil majors and the world petroleum industry; let us now turn to the structure of the Royal Dutch/Shell Groups whose subsidiary is studied in this work.

5.3 The Structural Organization of the Royal Dutch/ Shell Group of Companies

Figure 5.1:

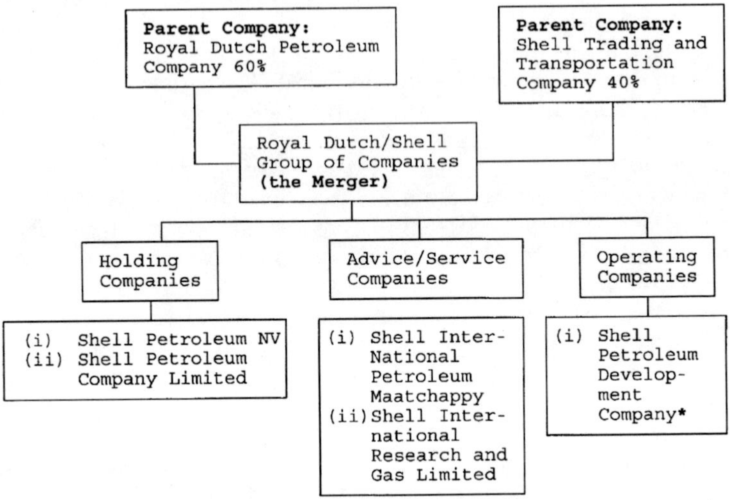

*This is the subsidiary of the Group that is operating in Nigeria.

As shown in the above diagram, the Royal Dutch/Shell Group of Companies consist of Holdings, Servicing and Operating Companies. The *Holding Companies* own shares between and among themselves, and in other associated companies. These are:

(i) Shell Petroleum NV; and

(ii) Shell Petroleum Company Limited.[25]

The *Servicing Companies* main business is to provide advice and services on a broad range of issues on oil, gas and chemicals to other associated companies operating in various parts of the world. They are:

(i) Shell International Petroleum Maatchappy BV

(ii) Shell International Mattscehappy BV (both of them are based at the Hague); and

(iii) Shell International Research and Gas Limited, Britain.[26]

It is important to note that these companies have, on their own, and through their subsidiaries, operated in the various regions of the world.

The *Operating Companies* explore and produce crude oil and natural gas. They also produce chemicals. They are wholly owned by the group with over 300 subsidiaries operating in various parts of the world. The Shell Petroleum Development Company in Nigeria and being the focus of this study, is one of the subsidiaries of the Operating companies[27].

The Royal Dutch/Shell Group, as diagrammatically shown above, is an established monopoly with a well vertically integrated network of operation. The Group is involved in all the stages of crude oil exploration and production with the *Specialized Companies* providing one kind of service or the other to associated companies within its fold. It is a close capitalist network of operation that ensures optimum use of both human resources and investment funds within and between all the companies under the Group. For instance, in 1978 when Gas Re-Injection Decree was promulgated, Shell Petroleum Development Company insisted that its sister company in Enugu would have to be invited for the gas preservation project. Its claim was that it has no specialty in that area. It did work out. This kind of service among the companies under the Group is common in all parts of the world where they operate[28].

Organizationally, the Group has centralized headquarters for all its companies which are based in Britain and Holland. The subsidiaries operating outside Europe merely receive orders from their home offices. For instance, the Group has two major research laboratories in Europe - one in the Hague and the other in London. Although the subsidiaries have laboratories, they all depend on the final results of research findings within the Group from the two laboratories[29].

Operationally, a great deal of the transaction that takes place between and among the members of the Group is to a large extent, intra-company reconciliations between the affiliates of the same mother company. Thus, the prices that are charged and paid for by the companies are not the result of distance trade; rather they are merely internal book-keeping prices or transfer prices involving various branches of the same corporate body[30]. For instance, the Group has one central refinery, about the largest in the world, located in Penis in the Netherlands. The bulk of the crude oil produced by the Group is shipped to this refinery. The shipment of the crude oil is usually handled by one of the affiliates of the Group, another one will have to ship the refined petroleum products to its petroleum pump stations located in various parts of the world. To a very significant degree therefore, from exploration through production to refining and marketing of crude oil and its by-products, it is an intra-company affairs[31].

So far, the structure of the Group has provided a clearer picture of its capitalist mode of operation. It has also brought into the fore, the location of the Shell Petroleum Development Company operating in the Niger Delta of Nigeria within the Group. Let us now discuss the structure of Shell in Nigeria.

Figure 5.2 The Structure of Shell Petroleum Development Company, (SPDC) Nigeria

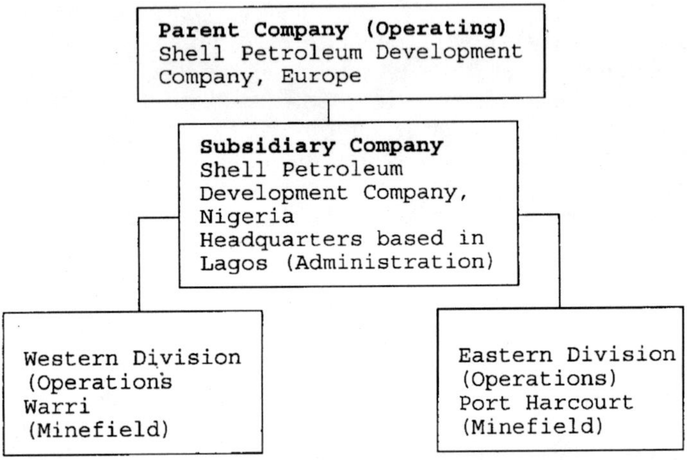

The structure of SPDC in Nigeria as shown in the above figure indicates that it is a subsidiary of Shell Petroleum Development Company, one of the Operating Companies of the Group based in Europe. SPDC is therefore, wholly owned by the parent company in Europe contrary to the belief by many a Nigerian, that the Nigerian government has equity shares in it. Rather, it is the Joint Venture Agreement in which the NNPC representing the Nigerian state, had 80% participating shares leaving Shell 20%. Since the state adjusted in line with the World Bank/IMF policies in 1986, the ratio of NNPC was reduced to 60% participating shares, and Shell's increased to 30%, and Agip took 5%, Elf with 5%[32]. It is the Joint Venture Shell services as its technical operator since 1973, that produces crude oil in Nigeria. Before then, it was Shell that operated the joint venture between BP and the company, an arrangement under which crude oil was produced then. More details on this in the course of the discussion.

As a subsidiary of the Operating Company of the Group, Shell's major business in the country is to explore and produce crude oil. The company's headquarters in Nigeria is located at Marina, Lagos. The Lagos office's basic function is administrative. The Managing Director of the company who is a European and normally posted from its parent company in Europe[33]. The company has two operational offices and both are located in the Niger Delta, its operational base or minefield. The basic function of the two offices is to explore and produce crude oil for use in the metropolitan countries. All activities of the company located in the east of the River Niger are under the control of the Eastern Divisional Manager with its base in Port Harcourt, Rivers State. Those in the west of the River Niger are under the control of the Western Divisional Manager with its base in Warri, Delta State[34]. The divisions has had more expatriate managers than Nigerians despite the fact the latter is not less qualified than former.

It is gleaned from the above that SPDC's structure in Nigeria is a reflection of its position in the Group. As a result, the bulk of its activities in Nigeria are concentrated in the Niger Delta. Perhaps, a knowledge of the background of SDPC in Nigeria might throw more light on its structure and operations.

5.4 Brief History of Shell Petroleum Development Company in the Niger Delta

To begin with, until 1956 when Shell struck crude oil in Oloibiri, very little was known of the true identity of the company in Nigeria. Its connection with BP was also shrouded in secrecy. It is important to first of all clarify the growing confusion on BP's connection with Shell in Nigeria in order to have a fuller understanding of the historical background of the later in the Niger Delta. The British government as earlier explained, had incorporated BP not only to break the monopoly of Royal Dutch/Shell Group in the oil and gas industry in the United Kingdom, but also, to ensure as well, regular supply of crude oil to its economy. It should be recalled that the BP catered for British interest in crude oil in the Persian Gulf. Little wonder it played similar role in Nigeria when Shell applied for permissions of Her Majesty in Britain to prospect for crude oil in the country, then a colony of Britain.[35]

The Mineral Act of 1914 had prohibited non-British companies from operating in its colonies as pointed out earlier. The Act was merely a ploy to guarantee regular supply of minerals including crude oil to metropolitan Britain, among others[36]. The Mineral Act also regulated the activities of Shell when it began operation in Nigeria. Thus, in 1936, BP merged interests with Royal Dutch/ Shell Group and incorporated Shell D'Arcy Oil Company. Shell

D'Arcy had previously represented the BP in the Persian Gulf and it was no surprise that the same company operated with Shell in Nigeria.[37] Although it was a joint venture between Shell and BP, BP's interest was mainly concerned with getting its share of crude oil produced, leaving Shell with actual exploration and production of crude oil. Their ratio in the joint venture was Shell/BP: 50:50[38]. Since Nigeria gained political independence 1960, and until 1979 when the Federal Government of Nigeria nationalized BP's shares in the joint venture in the country, the company's only interest was to collect its share of crude oil from the joint venture.

It can be argued therefore, that although there was the Shell-BP which dominated the Nigerian petroleum industry, it was Shell Petroleum Development Company, a subsidiary company of the Royal Dutch/Shell Group in Europe, that has continued to produce crude oil in the Niger Delta since 1956[39]. Now on the historical background of the SDPC proper.

The early origins of Shell in the Niger Delta can be traced to the colonial period. The reasons why Shell moved business to the Niger Delta among other parts of the Third World have already been explained that they need not detain us. Shell began oil exploration in the Niger Delta in the late 1930s, although an initial unsuccessful attempt was made by a German Company - Nigerian Bitumen Company in 1908 in Araromi in the present day Ondo state. In 1937, the company got its mining licence from the Crown Government in Britain, and conducted geological and geophysical surveys in that same year to delineate the oil bearing areas in the country. About 103,600 square kilometers in the Niger Delta basin was found to be oil-bearing which was then mapped out for intensive exploration. The outbreak of the World War I in 1939 however, disrupted further exploration by Shell[40].

By 1946, a year after the end of World War II, Shell returned to Nigeria to resume crude exploration. The company drilled the first well, but found no oil in 1951 at Ihuo, 16 km North-East of its base at Owerri in the present day Imo State. Two years later, the company drilled another well at Akata - 1, but found no commercial quantity of crude oil was struck.

By 1953, too, Shell arrived at Oloibiri town. Rather than move eastward of Oloibiri, that is, in direction of Nembe and the Atlantic Ocean, it went northwards since it had patches of dry land and plenty of monestry terrain. After three years of search for crude oil in the areas, it finally struck oil in commercial quantities a sport located 2 kilometers to the South of Otuabagi, 4 kilometers to the north of Otuogidi, 4 kilometers to the southwest of Opume and 5 kilometers further south of Emakalaka in the present-day Ogbia Council Area of Bayelsa State. The depth of the first well was 12,000 feet, and it produced at 3,000 barrels per day. Additional 11 oil wells were drilled in the Oloibiri field, but none was in Oloibiri town. It was the major find at the Oloibiri area other Western European oil companies got the assurance that the Niger Delta was an oil-producing area. As a result of the successful discovery of crude oil in the Oloibiri area, too, Shell D'Arcy Exploration Company changed its name to Shell- BP Petroleum Development Company, its parent company. Perhaps, the position of an official of Shell will be more instructive on this sudden change of name:

> It was more than a change of name. Shell worldwide is not associated with failure and it was because of the uncertainty of finding oil in the Niger Delta that made the company not to risk its name - Shell - at the initial stage of exploration[41].

Although the company's official might be right that Shell did not want to risk its name with failure in the oil industry, it is nothing less than one of the traits of the TNCs in the Third World. That is, they use different names in different countries to carry out the same business in order to hide their true identity in the host-countries. The sudden change in name obviously made it more difficult to know the actual identity of Shell[42]. It was with the new name that it then carried out further operations in the Niger Delta till 1979 when the BP's interest in their joint venture was nationalized.

Between 1956 and 1957 Shell further northward of the Oloibiri. By 1957, Shell drilled its first successful well in the Western Niger Delta at Erumukowharie - 4 kilometers from Ughelli in the present day Delta State, and about 100 kilometers to Oloibiri oil field. Major oil wells were drilled within the Ughelli communities. They included Afiesere, Umolo, Olomu, Uzeře, Oleh, Aviara, Olomoro, Kokori/Orogun oil wells. The company also found crude oil in other ares around Bomadi and Burutu[43]. By 1958, Shell's production output per day was 4000 barrels and, since this figure met its policy to begin the export of crude oil, the first shipment of crude oil to Europe from Nigeria was made in that year[44]. By implication, Nigeria became known as an oil-producing country. This intensified the pressure of European/American oil companies on the colonial state to gain entry into the Nigeria oil industry.

Thus, October 1960 when Nigeria gained political independence, and that broke the monopoly of Shell in the Delta area the post-colonial state amended few clauses in the Mineral Act of 1914 which allowed more oil companies other than those from Britain, to explore and exploit crude oil in the country. With the amended

Mineral Act in force in 1960, American, French, Italian and Japanese oil companies, among others, entered for the first time, into the Nigerian petroleum industry[45]. By 1961, subsidiaries of the American oil majors, notably, Mobil, Gulf and Texaco had been granted concessions and also begun oil exploration and production in the one-time Shell's exclusive[46].

As noted earlier in Chapters Two and Three, one of the hallmarks of the Nigerian political economy, is that the post-colonial state merely inherited the powers of the Mineral Ordinance of 1914 in spite of its amendment in 1960, leaving most of the clauses on the power of the state unaltered. Thus, the 1914 Mineral Act which was rephrased the Petroleum because of 1967 still rested the state like its predecessors, with the sole ownership of crude oil and recipient of rents, taxes and royalties accruing oil export and the oil companies. By implication, the state still reserved the power to grant and withdraw both prospecting and mining licences and concessional blocks to the oil companies[47]. Investing in the oil sector in the early years after independence was not really the priority of the Nigerian state. This was due to its perception of mining as a huge risk. The large capital outlay required, sophisticated nature of oil technology and highly skilled manpower needed in the oil industry, all of which the state lacked made it all the more so.

Granted that the reasons advanced by the state for not investing in the oil industry uptill the 1971 when it joined OPEC might be genuine considering the problems it was faced with as a new state. The same reasons however, can hardly justify why the state has not decisively intervened in the oil sector with a view to acquiring the oil technology and wresting control of the industry from the oil transnationals. Rather, the state has chosen to collaborate with Shell among other oil capital, albeit unequal relationship not unmindful of the consequences of its dependence on the oil companies and the pillage of its own oil-producing areas in the Niger Delta. Thus, what the Nigerian state has been able to do, is to remain a mere regulator of the country oil industry only in terms of passing laws most of which have sought accommodation with the oil companies.

One notable feature of the Nigerian economy uptill the 1970s, was the dominant and perhaps, monopoly position of the established large foreign companies which controlled virtually its operating in all sectors. To reverse the trend, Nigerian state promulgated the Enterprise Promotion Decree in 1972 as amended in 1977[48]. By the provisions of this Decree as amended, the oil producing companies were not significantly affected since oil exploration and production were and are still in the high technology area. Rather, the action of the Nigerian state towards the foreign oil companies was rooted in the OPEC's resolution which encouraged its member-locations to intervene significantly into their oil and gas industry. Coupled with the wave of indigenization, the Nigerian state introduced the joint venture agreement into its petroleum industry barely two years after joining OPEC.

Thus, in April 1973, the Nigerian state signed, for the first time, a participatory joint venture agreement with Shell in which it acquired 35 per cent. A year later, it was increased to 55 per cent. in 1979, the state increased its participation interest to 60 per cent leaving Shell with 40 per cent. By August 1979, BP's interest in the joint venture was nationalized because of the company's alleged business links with apartheid South Africa. With the nationalization of BP's shares, the state's interest increased to 80 per cent, leaving Shell with 20 per cent. However, in 1989, the Nigerian state had, in response to the demands of the World Bank and International Monetary Fund-led adjustment programme, divested itself 20 per cent of its interest in the joint venture with Shell. This was wholly offered to Shell to buy. But Shell acquired half of it which brought its participatory shares to 30% and invited both Agip and Elf to buy the reminder. This brought the new structure of the joint venture to NNPC 60%, Shell 30%, Agip 5% and Elf 5%. Table 5a for details. Shell, however, still remains the technical operator of the joint venture.

5.5. Shell/State Alliance in Oil Production in Delta Area

The task of this section is to examine broadly how Shell has, in conjunction with the state, operated in the Niger Delta. This is done in the context of the three social forces directly involved in crude oil production namely the state, the foreign oil capital (Shell in this case) and labour (workers). The state, for instance, formulates the policies and creates the political and business environments that are conducive for capitalist development in the oil industry and the host areas as well. Shell provides the technology for crude oil production, and in most cases, the capital as well. The workers supply the needed labour for the successful production of crude oil. To gain a further knowledge of Shell's capitalist exploitation in the oil industry and the Niger Delta each of them is treated separately.

5.5.1 The State Laws on the Oil Industry

To a very significant degree, the nature of the colonial and post-colonial Nigerian state as earlier explained in Chapter Three, remains structurally the same[49]. The broad spectrum of laws and

Table 5.1: Government Participation Interests in The Oil Companies

Company	Participation %	Date Acquired	No. of Oml/Opls	Production (Barrel/ Day)
Elf	35	1-4-71	4	40,000
	55	1-4-74	4	85,000
	60	1-4-79	4	78,000
Agip/Phillips	33½	1-4-71	4	30,000
	55	1-4-74	4	125,000
	60	1-4-79	4	230,000
Shell/BP	35	1-4-73	58	1,240,000
	55	1-4-74	58	1,380,000
Shell	60	1-7-79	58	1,360,000
	80	1-8-79	58	1,360,000
	60	1989*	58	n.a
Gulf	35	1-4-73	10	368,000
	55	1-4-74	16	390,000
	60	1-7-79	16	380,000
Mobil	35	1-4-73	4	210,000
	55	1-4-74	4	245,000
	60	1-4-79	4	380,000
Texaco	55	1-5-75	6	10,282
	60	1-7-79	6	56,000
Pan Ocean	55	1-1-78	1	10,000
	60	1-1-79	1	10,000
Ashland	Production Sharing	June 1973	2	13,000
Agip Africa	Service Contract	Sept. 1979	6	n.a.
Elf Acquitaine	Service Contract	Sept. 1979	3	n.a.

Source: NNPC Joint Venture Department

n.a. not available

*As a result of IMF and World Bank directive to the Nigerian state since it adjusted in 1986, it divested 20% in 1989.

It is important at this point to discuss Shell's mode of operation in the Niger Delta.

actions of the Nigerian state on the country's oil industry therefore, have been no less than the extension of the 1914 Mineral Act which vested all powers over resources (solid and liquid) on the state. This is because nothing has changed in the Act except the replacement of the clause 'Crown Government' with 'Nigerian Government' in addition to the amendment which now permits other European and American oil companies to do business in the country's petroleum sector. For instance, the post-colonial state like its predecessor, has the exclusive right to issue mineral prospecting and mining licences. It also had the sole power to the collection of royalties, rents and fees from the oil companies[50].

Other the state's laws governing the country's industry oil industry were drawn extensively from existing bodies of international laws without putting in place, the necessary instruments for their enforcement locally. Owing to the peculiar nature of crude oil production and transportation, for instance, the colonial state enacted the Oil Pipeline Act in 1958. By this Act, oil companies were to obtain a legal permit that would empower them to lay oil pipelines linking up their various oil wells and fields across the Niger Delta, which, were then connected to their terminals from where crude oil is finally exported. The major role of this Act was to boost the revenue base of the state[51]. However, apart from the revenue that accrue to the state from the licences issued to the oil companies, the Act as it implies in the US and European oil industries is to present the oil companies from willingly encroaching on other parcel of land and operate with environmental consciousness. But in Nigeria, the state merely continued with policy after 1960 ostensibly with how much revenue it can generate. Using it partly as a safeguard against environment protection was out of it, which then gave Shell the leeway to lay its pipelines across sacred sites and homes of the people who live in the Niger Delta[52]. Other legislations like the Mineral Oil (Safety) Regulations of 1963, compelled Shell, among other foreign oil companies operating in Nigeria to abide by the ethics of good oil field practice. That is, it prohibits the operating oil companies from discharging noxious gases into the environment or face penalties for any violation. The Nigerian state has not been able to enforce the Act it pollutes by virtue of the legal implication of its participatory share in the joint venture agreement with the foreign oil Companies[53]. Case histories of Shell's blatant violation of this Act in its operational base in the Niger Delta are discussed in great details in subsequent chapters corroborates Eboe Hutchful's submission elsewhere that, because the Nigerian state relies largely on the oil companies for its revenue, it hardly implements its laws on the industry especially as they affect the oil companies[54].

The state's Oil in Navigable Waters Act of 1968, like the Mineral Oil Safety Act of 1963, was adopted from the Sea by Oil Act of 1954 as amended in 1962 Geneva. This Act prohibits the oil companies operating in Nigeria from discharging oil or any mixture containing oil into the territorial navigable waters from land[55]. In fact, it is in part expected to protect the environment. Yet, Shell still dumps harmful and untreated drilling wastes into the environment, and spills crude oil at the terminal, from burst hoose. The state own inspectorate agency, The Petroleum Resources Department, does not have the equipment for monitoring the activities of Shell, making it all easy for the company to be reckless. In other words, should the Department decide to enforce the Act, it would have to rely on Shell to obtain the volume of oil spilled, and the magnitude of pollution[56]. Shell and other oil companies. The long-drawn bureaucratic procedure that any culprit of this Act has to undergo, even worsened the chances of its enforcement. For instance, the suspect is first subjected to the bureaucracy of section 12 (i) (c) of the Interpretation Act of 1964 and if found guilty, fined one hundred naira only (N100.00) or imprisonment of six months or both.[57] For the period covered in the study, there was no instance the state had instituted a case against Shell in spite of the company's glaring violation of the Act as evidenced in its pollution of the Niger Delta[58].

More than any previous state legislations on the country's oil industry, the Petroleum Decree of 1969 was about the most comprehensive particularly in terms of its scope and issues. The Decree vested the full ownership and control of petroleum under or upon all land in the country, its territorial waters including its continental shelf in the Nigerian state. By the provisions of the Decree, the Minister of Petroleum Resources was only empowered by the Nigerian governments to issue a licence or grant a lease to an operator over an expanse of land to prospect for, drill and extract crude oil. The Decree also provides that, it is mandatory for the licencee or oil lease to pay adequate compensation for its use of private land or any damage done to it. The decree made it explicitly clear to the operating oil companies that they should abide by the ethics of good oil field practice especially in protecting the environment of the host-communities where crude oil exploration, exploitation and transportation are currently done[59]. Yet, the state has been unable to control the activities of the oil companies in the Delta area in spite of the decree. Why? It is partly because the Petroleum Decree of 1969 is not significantly different from the 1914 Mineral Act in content since the state's overriding interest is how much fund it can raise from the operations of the petroleum companies, but not to protect the environment of the host communities and care for the well-being of the people who live there. The fact that the state is itself, a pollutant makes it all the more so. All is to partly inform the basis of the rentier thesis on the Nigerian state by scholars like Forrest, Ihonvbere and Turner.

One other area where state legislations on the country's oil industry have also sought accommodation with the oil companies is land. Shell, like other oil companies, needs land for most of its operations. For instance, right from exploration through production to transportation of crude oil by pipeline, land is needed. Large hectares of land is also needed for the construction of oil terminals and 'tank-farms'. Land is required for petroleum depots, flow stations, flare sites, offices and staff quarters. These facilities will only be efficiently operated with good access roads, which again, requires land. Land is also needed for the construction of airstrips to ease the transportation of personnel and oil tools to the site of operation. All this informs the content of the Land Use Decree of 1978, a decree that vested sole ownership of all lands where oil is explored, produced, transported and stored in the state. It also defines the contents of land in the country[60].

On account of the materials interest of the state in the oil industry therefore, the Land Use Decree provided a buffer for the oil companies to be more reckless in their operations. This is part explains why Shell can afford to flare gas day and night since 1956 when it found oil in the Niger Delta. The same is also true why Shell is protected by the state over litigations on land where oil is explored or produced.

Oil Pipeline Act of 1956, companies for an oil-pipeline right-of-way for 99 years over the surface of a stretch of land under which pipes conveying crude oil, or gas to flow stations, then the terminals, were laid[61]. What is important to the state was the revenue it (state) could generate from the issuance of the licence. Not surprisingly, the Act was continued with in the post-colonial period as the Oil Pipe Line Act of 1963[62]. Lease period was later reduced to 60 years and later to 20 years. The implication of the Oil Right-of-Way Act of 1965 for the host oil areas is that, land that falls under the right-of-way of the oil companies would not be compensated for; and it does not matter what kind of land[63]. Case histories of Shell's right-of-way running through villages and sacred bushes in the Niger Delta and with grave consequences for the culture of the people who live there abound. Because the law also empowered Shell among other oil companies, to lay its pipes on, over and under land, and as well exercise control over the surface of the specified it pipes pose hazards to the inhabitants of the area as they cross manifold of oil/gas pipelines under high pressure in their daily chores.

The Oil Right-of-Way Act of 1965 did not confer the right or certificate of occupancy on Shell. This means that the land on which it operates still belongs to the state by virtue of the Petroleum Decree of 1969. Under this Decree, the community where land is found still has right to land but not the oil produced. So, oil companies paid some compensations. To circumvent the impasses brought about by the Decree the Land Use Decree was enacted in 1978. This is not an assumption that the Land Use Decree was enacted purposely to protect the petroleum companies, but to a large extent, it reduced the friction that Shell and other oil companies have had with members of their host-communities in the Niger Delta since the 1950s. Another main feature of the Land Use Decree of 1978 is that, the State Government is empowered to revoke the right of occupancy of any person for the overriding 'interest of the public', which is no less than the interest of the state managers. It (state) also, has power under S.47(b) of the same Decree, to grant statutory right of occupancy to Shell and other oil company as the case may be.[64]

In fact, no piece of state legislation on the country's industry has so eroded the right of the members of the oil-communities over their farmland than the Land Use Decree of 1978. For, it turned the inhabitants of the oil-bearing areas into squatters in their ancestral homes. With this decree, the state has not only sought accommodation with Shell, among other oil companies, but also, denied the people who live in its host communities most their farmlands thereby causing land shortage in the area. A detailed discussion on this is taken in chapter seven.

5.5.2 Shell's Mode of Producing Crude Oil[65]

5.5.2.1. The Oil Servicing Companies (OSCs)

One of the after-effects of the Anti-Trust Law of 1911 enacted by the government of the USA, is the emergence of independent and small capital-based, but very specialized oil-servicing companies (OSCs). Hitherto, they operated either as branches of the oil monopolies or as fully incorporated companies operating on minimal scale. Though the Anti-Trust Act gave rise to many OSCs, the urgent need to specialize in one of the stages of the oil industry and in the production of crude oil reduced as raw materials are being sourced in the countries of the Third World whose business environment is unpredictable. Thus, by the time Shell found oil in the Delta area in 1956, the production of crude oil was so specialized that to cement an oil well, that is ready for production, no less than 10 specialized oil servicing companies would be involved. This explains why Shell still engages the services of the OSCs in its operations in the Niger Delta area and elsewhere in the world. A little elaboration on this might throw more light on some aspect of its capitalist operation in the Delta region and the implications for the host country.

First, Shell starts business by obtaining the relevant exploration and mining licences from the Nigerian state. At the exploration stage, which includes, among others, geological/geo-chemical surveys, geophysical and seismic operations, Shell contracts the jobs to the OSCs, virtually all of which are foreign. One of the major contractors for Shell at this stage, is Seismograph Services Limited with both American and British origins.[66] At seismic level, the contractor company does all the job with its own engineers and casual workers with a few of Shell's field staff who occasionally came in to ensure that the standards of operation are met. The problems of environmental pollution actually begins at this stage. Yet, no consideration is given to the treatment of the harmful waste generated in the process by both the state and Shell including its OSCs. For instance, a lot of the chemicals and explosives used by the OCSs, general harmful waste that are deposited on the environment, thereby posing danger to aquatic organisms, vegetation and even human beings. In fact, the OSCs operate with utmost disregards to the environment of the oil communities since they are merely contractors to Shell which the state and the people who live in the host areas recognize as the major oil operator. In other words, they owe no obligation to the state, Shell and the people. In any case, the OSCs would not have operated otherwise if they do[67].

Next to seismic operations, is drilling, in which no less than three oil servicing companies are involved. They include the drillers, the marketers of the drilling chemicals/equipment and those that service the equipment/rigs and perform other engineering functions. Forex, Anadrill and Key Drill are among the OSCs that drill for Shell. Bariod and Dressing Nigeria Limited supply drilling equipment and provide engineering services respectively. Schlumberger, Flopetrol among other OSCs, service the company's oil platform[68]. The damage done to the environment emanates from the wastes generates from drilling which are all emptied on it. Details on this are discussed in the next three chapters.

When an oil well is finally prepared and ready for production, Shell still needs the services of other OSCs for the supply and laying of oil pipelines, which are connected to the flow stations, then the terminals. Otis is one of the OSCs that does this job for Shell. Solus Schall provides diving services for Shell, while Niger Cat undertakes most of the construction of its access roads. In fact, Shell contracts virtually all its job, to the OSCs except the supervision of production and export of its own share of crude oil from the joint venture. Which means that Shell is indeed, a contractor. It is not to assume that Shell cannot handle its operation. Of course, it can. Rather, the rule of capitalism, not lease at its monopoly stage, compels Shell to hire the best services from the best company aimed at enjoying economies of scale. But it is in its mode of operation that the underdevelopment of the Delta can be partly located.[69]

The implications of Shell's mode of operation of crude oil on the economy of the Delta area in particular and Nigeria generally are numerous and with grave negative consequences. First, when Shell contracts job to an oil servicing company, it pays directly to the company. How the company does its jobs, recruits labour, among other issues, do not concern Shell any longer. On its part, the OSC brings on its highly skilled staff hires few locally bring in most of its equipment including vehicles from its home country it requires there, is to hire foremen who then recruit casual workers for its operation. This mode of operation by both Shell its OSCs particularly with respect to labour exploitation has heightened unemployment rate in the Delta area and Nigeria generally. The pertinent question is: How can Shell and its contractor companies introduce contract labour into a country where secondary school leavers and of Universities and Polytechnics with relevant disciplines for the oil companies, have very few of them employed on contract? This brings into the fore, the nature of the Nigerian state, its approach to labour, and alliance with foreign capital to further its selfish economic interest. Elsewhere in Africa, this can hardly happen in an oil sector. In Libya, for instance, which is an Africa oil-producing country, the OSCs are under the state's law on labour compelled to raise the bulk of their technicians and tool pushers not only from within the country but also on permanent and pensionable basis. Not only that, that OSCs will have to seek the approval of the state before firing an indigene in their payroll for whatever reason[70]. Suffice it to note that, if the OSCs fire workers at will in Nigeria, it is an indication of the weakness of the Nigerian state and its indifference to the development of the oil industry. Herein lies its role in the underdevelopment of its oil-bearing areas.

Second and final, is the influx of the foreign OSCs into the Nigerian petroleum industry which has been detrimented to the growth of local businessmen in that sector. in the area. Beginning from the stage of exploration through production to the transportation of the crude oil to the terminals from where it is exported, it is all dominated by the major foreign oil companies and the contractor service companies. Although the high-level technology required in these jobs explains in part why the OSCs recruit most of their highly skilled manpower from their home countries, and very few locally, the inability of the state to come up with a stringent policy on the operations of the companies makes it all the more so. Unlike Nigeria, both Algeria and Libya, two major African oil-producing countries, have clear-cut policies on their petroleum sectors. For instance, through proper training at both local and overseas levels, they have developed the local staff that now manage most of the critical stages of their petroleum industries. Others who are not in the state employment but have earlier benefitted from the state's manpower development programmes now compete favourably with the OSCs. Their relatively developed steel sector made the indegenization of oil technology possible as oil papers, values and tools are fabricated locally. While Libya has been beset with religious crisis since late 1980s, which shattered its economy[71].

This is in sharp contrast with Nigeria where the jobs reserved for the local businessmen are basically catering, gardening and other menial jobs.[72] Part of the reason why it is so in Nigeria, is because the state is not really interested in developing its local productive force and its economy. For instance, the indigenization decrees of 1977 placed the oil industry under the high-technology category and left it as the preserve of the oil companies[73]. Worst still, it began steel development with the hope of reversing the country's technological dependence on foreign capital, but to deepen it. Above all, the state is unwilling to intervene decisively in oil production since most of its custodians are board members in the OSCs operating in the country. So, the influx of the OSCs into the Nigerian petroleum industry benefits those in charge of the state. While the state has discouraged the emergence of a strong local business class, this class itself is even more comfortable with being petty contractors to the OSCs. More details on this later.

5.5.2.2 The NNPC/Shell Joint Venture

Many are often misled by the ratio of the Nigerian state interest in its joint venture with Shell to believe that the company is no longer a transnational corporation since it, through the NNPC owns over three-quarters of the shares in the agreement wrong[74]. As earlier explained, Shell Petroleum Development Company is wholly owned by the Shell Group based in Europe. The joint venture is an agreement signed by both the state and SPDC on how to produce crude oil in the country. Legally, it is the joint venture, which is distinct from SPDC as a company, that has been producing crude oil for the partners since 1973. Shell is only an operator of the joint venture. It is all part of the plots of Shell to domesticate the Nigerian state. Even the state managers so conduce to such lopsided legal agreement because it benefits them. The ratio in the joint venture as shown earlier, indicates how much each partner is to contribute to the cost of producing crude oil and its share of the crude as well. In 1973, it was NNPC/Shell 35:65; in 1974, 60:40; in 1974; and 80:20 by August 1979. In 1989, it was NNPC/Shell/Agip/Elf: 60:30:5:5.[75]

Although both partners that is, NNPC/Shell for the purpose of this study, are to contribute funds for carrying the exploration and production of crude oil, in proportion to their ratios, the state has hardly been forthcoming with its cash call[76]. In most cases, Shell contributes both funds and equipment for the production of crude oil by the joint venture. Not only that, Shell also provides the necessary manpower and equipment with which crude oil is produced. All cost incurred by Shell as the technical operator of the joint venture right from exploration through production to the terminal is calculated in crude oil and then its equipment subtracted from the total quantity of crude oil produced; the remainder is then shared by the partners in their ratios in the agreement. Shell exports its share of the crude oil to Europe. The state refines part of its share of the crude oil and exports the rest.

One of the consequences of the manner in which the NNPC/Shell joint venture is operated, is that oil technology will for a long time to come, remain with Shell or any other foreign oil operating in Nigeria. Those in charge of the Nigerian state are aware of this and other negative consequences of a lopsided joint venture, but prefers it done that way because they benefit materially from the process. Herein lies part of the irony of state capitalism.

The Nigerian state experience with the operation of joint venture contrasts with Nigeria. In 1956, oil was discovered in Algeria, the same year oil was found in Nigeria Algeria, like Nigeria, introduced joint venture agreement into the upstream sector of its oil industry. In 1963, the Algerian state set up a public oil corporation, Sonatrach, to manage the country's petroleum industry.[77] Nigeria did the same by the setting up the Nigeria

National Oil Corporation (NNOC) in 1971. The NNOC was dissolved in 1977 for inefficiency and, in that same year, established the Nigerian National Petroleum Corporation (NNPC) to carry out similar functions particularly that of overseeing the country's petroleum industry on behalf of the state. By 1966, Sonatrach had signed joint venture agreements with the foreign oil-companies operating the country's oil sector. In 1971, its counterpart in Nigeria (first by NNPC and as from 1977, by NNPC) signed one form of joint venture agreement or the other with foreign-oil companies operating in the country. It was however, in 1973 that the NNPC signed joint venture agreement with Shell[78].

As of the late 1980s, Algeria was far ahead of Nigeria in terms of the development of its internal technical capacity to find, win and refine crude oil, independent of the foreign oil companies. Unlike in Nigeria where virtually all the gas produced in the process of producing crude oil is flared, Nigeria was and still able to export its gas. For instance, by 1974, that is, eight years after Sonatrach had signed joint venture with the oil companies, it had indigenized oil technology as well as acquired the capacity to produce crude oil and refine it as well. To achieve this feat the state had to put the country's iron and steel industry on stream, and ensured that local inputs such as drilling pipes, and values were produced locally. By 1976, the oil companies found themselves almost redundant as Sonatrach had taken control of virtually all aspect of the Algerian petroleum industry. The Algerian situation would have been ideal for Nigeria in the 1980s. However, by the turn of 1990s when the country began to face serious political, economic and social problems, a lot of the gains made in the country's oil industry were consumed by the crisis[79]. But all the same, Nigeria still lags far behind Algeria in technology in the petroleum industry. For, the Nigerian experience is such that the NNPC, as official of the corporation has aptly summed it, "is what the state wants it to look like - an instrument for capital accumulation by those who manage the state".[80]

5.5.3 Contract Labour

As earlier explained, colonial capitalism created a pool of urban-base unskilled labour from which capital drew its casual workers. The monetarization of the colonial economy made it possible for labour to sell its talent for whatever wage that capital was willing to pay. All this was continued with in the post-colonial period, which explains why those entrusted with the management of the Nigerian state were basically interested in creating conducive atmosphere that permits state-transnational capitalism to flourish. Rather than formulate labour policies that would make it difficult for foreign oil capital, not the least Shell, to hire and fire local labour willingly, it conduces to its policies. The state's anti-labour policy stance in the oil industry and other sectors of the economy, made it all the more possible for Shell and its contractor companies (OSCs) to hire and fire workers at will. The introduction of contract labour might be aimed at minimizing labour unrest, and reduce overheard costs.[82] It means more than that; for, under contract labour, Shell and its OSCs, do not regard such categories of workers as full staff of the companies. Hence they are not pensionable, no medical allowance, no salary increase and no union action.[83] Suffice it to say that, Shell uses contract labour to domesticate labour and exploit it as well in the oil industry. While it has heightened unemployment in the economy of Nigeria inclusive of the Delta region, it enabled the oil companies to provide jobs for their home countries in the country. For instance, under the pretext of contract labour, expatriate staff are hired by the oil companies to fill vacancies in the country that Nigerians are qualified to occupy.

So far, the Shell/state alliance in the production of crude oil has given us an insight into the forces of state transnational capitalism have combined to pillage the Niger Delta. The next three chapters discuss the consequences of all this on the economy of the Niger Delta area.

Notes and References

1. Anthony Sampson's work was based on his field experience as an energy journalist in the late 1950s. His orientation as a journalist coloured his work and to that extent, he failed to analyze the oil companies as corporate finance capitals. In the process, he missed the root causes of the crisis of the petroleum sector. For details, see Anthony Sampson, *The Seven Sisters: The Great Oil Company and The World They Made* (London: Hodder & Stonghton, 1975). See also, Louis Turner, *Oil Companies in the International System*, (London: George Allen & Unwin, 1976). For further details, see Peter Norre and Terisa Turner (ed) *Oil and Class Struggle* (London: Zed, 1980) and particularly Chapter Two of the work. See Julius Ihonvbere, 'Labour State and Capital in Nigeria's Oil Industry' (forthcoming book); Susane Cronje, *The World and Nigeria: The Diplomatic History of the Biafran Civil War 1967 - 1970* (London: Sidnwick and Jackson, 1972) and various of Terisa Turner on the Nigerian petroleum sector. See also Michael Tanzer's *The Political Economy of International Oil*

Companies and the Underdevelopment Countries (London: Temple-Smith, 1969).

2. See Louis Turner *ibid*; Norre and Turner *ibid*, and Tanzer *ibid*. Among the individual entrepreneurs who pioneered the oil business were Rockefeller - an American; Marcus Samuel, a Briton. Both of them once worked in the oil companies; Henri Deterding was a Dutch merchant; while William Knox D'Arcy was a British merchant.
3. For the early history of the origins of the oil companies, see for details, Sampson, *op. cit.* and Turner op.cit. For further details, see also, M.O. Feyide, 'Oil in World Politics, paper presented at the J.I.C. Taylor Memorial Lecture for 1986, University of Lagos, 13th-14th March, 1986.
4. See Sampson *ibid.* and Turner *ibid.*
5. Cited in Sampson, *ibid.*
6. M.O. Feyide, 'Oil in World Politics' *op. cit.* pp.23-24.
7. See in particular, Chapter Two of Louis Turner, *Oil Companies in the International System*, *op. cit.*
8. See *The Story of Shell*, *op. cit.*
9. M.D. Feyide, *op. cit.* See also, Lenin's *Imperialism, The Highest State of Capitalism*, *op. cit.*
10. Turner, *op. cit.* Feyide, *ibid.*
11. It was difficult getting the actual percentage point of companies had after the 1911 Anti-Trust Law. This is because the officials of some of the offspring companies of the Rockefeller Trust particularly Mobil which operates in Nigeria does not even know.As for Shell, the rule of the game is secrecy. But for some officials (who wanted to be unnamed) disclosed that neither Marcus Samuel nor Henri Deterding the founders of Shell Transport and Trading Company and the Royal Dutch Oil Company respectively had shares in the present structure by the companies. In the light of the above problems, it was difficult having an accurate number of equity shares held by the shareholders.
12. See Louis Turner's *Oil Companies ... op.cit.*
13. *ibid*; see Sampson, The Seven Sisters, *op. cit.*
14. *ibid*, (Turner).
15. *ibid.*
16. *ibid.*
17. *ibid.*
18. I drew heavily on Turner's work in this section. See his *Oil Companies ... ibid*, pp.38-67.
19. *ibid.*
20. *ibid*, but see also, *The Story of Shell*, *op. cit.*

21. So much work has been done since the continent of Africa was partitioned in the late 19th century. See for instance, the various works of Claude Ake particularly his *A Political Economy of Africa, op. cit.* Earnest Brett, *Colonialism and Underdevelopment in East Africa ..., op. cit.*, Walter Rodney's *How Europe Underdeveloped Africa, op. cit*, Carl Widstrand *et al*, *Multinational Firms in Africa op. cit.*, *Muller et al*, ed. *Africa Undermined ..., op. cit.*
22. See Turner, *op. cit*; and Sampson *op. cit.*
23. *ibid*, (Turner).
24. *ibid.*
25. *The Story of Shell op. cit*; also see, Shell Information handbook, *op. cit*, pp.7-13.
26. *ibid.*
27. *ibid.*
28. Interviews with some of the officials of the Joint Venture Department of NNPC in Lagos in August 1990 mid 1995.
29. Interviews with some of the officials of Shell in Lagos in January, 1991, and later in 1995.
30. *ibid*, for details on the operation of the transnational corporation, see Bipled Dasgupta's 'The Changing Role of the Major International Oil Firms' in Carl Widstrand *et. al. The Multinational Firms in Africa ..., op. cit.*
31. See note 29 above.
32. This is the current position of the Joint Venture between NNPC, Shell, Agip and Elf. It is not clear from NNPC whether this ratio has been ratified as of 1996.
33. See note 29 above.
34. *ibid.*
35. Based on the author's interviews with officials of Shell (top management) in Lagos in June 1989 and later in 1991 and 1994.
36. *Ibid.*
37. *Ibid.*
38. *Ibid.*
39. *Ibid.*
40. See note 35 above. It was John Simon Bergheim that formed the Nigeria Bitumen Company in 1906. For further details, see Chapter 7 of John Carland's *The Colonial Office and Nigeria 1898 - 1914* (London: Sandford University Press).
41. Based on author's interviews with some top management of Shell in Lagos in June 1989 and 1992.

42. Changing identities of the TNCs is one of the attacks leveled against them in the world economy by the UDT. See, for instance, Carl Widstrand *et. al. Multinationals Firms in Africa op. cit.*
43. See note 35 above.
44. *ibid.*
45. *ibid.*
46. *ibid.*
47. The 1969 Petroleum Decree which became an Act for instance, contained initially all the clauses of the 1914, 1959 and 1963 Minerals Act. See the Petroleum Decree No.51 of 1969 (Lagos: Federal Ministry of Information, 1969).
48. See the Nigerian Enterprises Promotion Decree of 1972 as contained in the Federal Military Government of Nigeria Supplement to Official Gazette Extra Ordinary Vol. 58 February 28, 1972.
49. For more details on the character of the Nigerian state, see Chapter Two of this work.
50. The 1914 Mineral Act as amended in 1959 and 1963 was finally incorporated into the *1969 Petroleum Decree (Act) op. cit.*
51. Cited in V.O. Jonah 'Machinery for Implementing Environmental Pollution Laws in Oil and Gas Operations' in *The Petroleum Industry and the Nigerian Environment op. cit.* 1985. More details of the 1958 Act as amended in 1965 are all contained in the article and the sections on Law in the 1981 and 1983 issues of the Conference proceedings.
52. *ibid*, more information was obtained from the author's various interviews with some officials of Shell who are indigenes of the oil-producing areas in the Niger Delta.
53. See not 51 above.
54. See Eboe Hutchful's chapter in Claude Ake's ed. *Political Economy of Nigeria op. cit.*
55. See note 51 above.
56. Based on the author's interviews with some of the officials of the Petroleum Resources Department in 1990, 1992 and 1995.
57. See note 51 above.
58. *ibid*; Based on author's interview with some officials of Shell's Legal Department, and NNPC Legal Department, Lagos.
59. See the Petroleum Act of 1969 *op. cit.*
60. Based on the author's field trips to some of Shell's areas of operation in the Niger Delta in 1989, 1990, 1991, 1995 and 1996.
61. Laws of Nigeria; Oil Pipeline Act of 1956.
62. *ibid*, as amended in 1963.

63. Laws of Nigeria: Oil Right-Way Act of 1965.
64. Land Use Decree of 1978.
65. This Section is based mainly on the researcher's several interviews with both office and field staff of Shell and other oil companies/OSCs operating in Nigeria and elsewhere in Europe and America.
66. Shell is always looking for the best of the OSC's and to achieve this, it calls for tenders from these companies which are mainly foreign. This has made it difficult to have a list of permanent contractor oil companies to Shell. However, names of some of the OSCs are obtained during the researcher's field trips to Shell, NNPC, Ministry of Petroleum Resources.
67. Based on the author's interviews with some of the drilling engineers in the oil companies at Warri, 1990.
68. See note 66 above.
69. *ibid.*
70. Based on the author's interviews with officials of one of the OSC's operating in Libya, Lagos, 1989, and Austin, Texas, 1996.
71. *Ibid.* See Terisa Turner *et al*, ed. *op. cit.*
72. Based on the author's several field trips to the Niger Delta area between 1989 and 1997.
73. See the section on Political Economy of Nigeria in Chapter Two of other's work.
74. In most of the interviews that the researcher had with some members of the academic even particularly those in related disciplines like Political Science, International Relations, Economics, there is a total misconception of what Joint Venture means. One of the members of the academic even challenged me that Shell is owned by the Federal Government. I had to educate most of my respondents on the meaning of the joint venture and how it works in the country's oil industry.
75. The current figure of the ratio was announced by the Federal Government in 1989 and it divested according to IMF and World Bank instructions. The researcher confirmed the ratio from officials of Shell and those of the Joint Venture Department of the NNPC, but it was yet to be ratified as of 1996.
76. Based on the author's interviews with officials of Shell and those of the joint venture unit of NNPC in Lagos, 1989. For more details, see J.O. Orife, 'The Role of Joint Ventures as the co-ordinator of Oil Industry Activity', paper presented at NNPC Public Relations Seminar, Lagos, 1987; and *Government Joint Venture Participation Interests in the Oil Industry (mimeo)* (Lagos, NNPC, 1986).
77. See Peter Norre and Terisa Turner ed., *Oil and Class Struggle op. cit.*

78. See Chapter Five of Norre and Turner ed., *Oil and Class Struggle op. cit.*
79. *Ibid.*
80. Based on the author's interviews with officials of the joint venture unit of NNPC. In fact, this view is also held by many Nigerians working in Shell.
81. For more details, see Marx's 'Theory of Surplus Value' as explained in Dan Nabudere's *The Political Economy of Imperialism op. cit.*
82. Officials of NUPENG have blamed the use of contract labour by the oil companies on the weakness of the Nigerian state and its anti-labour policy stance The researcher also interviewed an official of one of the OSCs in Libya on the issue of contract labour. The respondent said, 'no oil company operating in Libya fires a Libyan staff with the consent of the government'.
83. Based on the author's interviews with some of the foremen in the Delta Niger in 1990 and 1995.

Chapter 6

Shell, Environmental Degradation and Rural Poverty in Its Host Oil-Producing Areas in the Niger Delta

6.1. Introduction

As noted, Shell controls over 70 percent of the oil activities in the Niger Delta area. This explains in part why the company is most implicated in incidents of environmental pollution of, and incessant agitations by the people living in the oil-rich Niger Delta. However, other operating foreign oil companies, namely, Chevron, Mobil, Texaco, Agip, Pan Ocean and Elf are no less guilty than Shell except that their scale of recklessness is relatively smaller. Suffice it to note that, the reckless abandon in which oil is produced, they, like Shell, degrade the environment of their host communities across the Niger Delta. As its primary object, therefore, this Chapter examines the general trend of the environmental, and socio-economic consequences arising from the operations of Shell for its host oil producing areas in the Niger Delta region, taking note of those of other foreign oil companies operating in the area. Within this context, it particularly critically looks into how the degradation of the environment of the oil-bearing areas by the oil company, has come to shape the nature of poverty of the people who live in its host rural oil communities. It analyses the efforts made by the inhabitants who play host to its operation, to develop alternative strategies of survival, how they have reacted to their exploitation noting the actors and actresses, and institutions involved, the interests they represent, and their strengths and weaknesses. It, finally, examines the responses of both Shell and the Nigerian state to the sad plight of the oil producing communities.

6.2. Environmental Degradation and Rural Poverty

To begin with, there is an intricate (linkage between the relative reckless manner in which crude oil is explored and produced by Shell, the degradation of the environment of its host communities, and the contemporary increasing incapacity of the people who live in these areas to reproduce themselves socio-economically.

This is all the more because as a non-renewable natural resource, co-existing with other environmental resources, the process of finding and winning crude oil inevitably results in the removal of other natural resources from the earth crust, which, in turn, upturns its balance, and degrades the larger environment. Also, crude oil production causes the forceful wastage of other natural resources such as natural gas that is either associated with it, or land, which is an intricate component of the environmental resource. One of the main implications of all this, is that, apart from the environmental degradation arising from the manner Shell operates other natural resources are being incapacitated from renewing themselves and at their own pace if they do at all. Human beings have been particularly badly hit in the process since most of their economic activities tied in one way or the other to the environment, are either disrupted or destroyed, or even both.

In our own universe of discourse therefore, the environment is used to mean the entire environmental resources of the Niger Delta, including the culture of its people and other natural things attached to it, and how they are used and managed to support the capacity of the people who live there, to reproduce themselves materially and spiritually. Environmental degradation means the *inability* of the environmental resources, like the atmosphere, rivers, soil, vegetation to renew themselves naturally, which, in turn, incapacitates the people of the Niger Delta from regenerating themselves, socio-economically. Rural poverty, in the context it is used here, means a situation of severe deprivation suffered by the people living in oil-producing areas in the Niger Delta region, and at a level that their minimum needs of life can scarcely be met even at the lowest standard necessary for their survival particularly in terms of nutrition, housing, health and cultural life. The reality is that, the majority of the people who live in the oil-producing areas of the Niger Delta where over 90 percent of the country's wealth is currently produced, are poor, malnourished, miserable, and perpetually vulnerable. It is by no means an assumption that, rural poverty did not pre-date the introduction of colonial capitalism in the Niger Delta as explained earlier. At issue, rather, is the nature of rural poverty in the host oil- bearing areas occasioned by the operations of the foreign oil companies like Shell, Chevron, Mobil. That which has left the people particularly increasingly incapable of improving their material and spiritual well-being.

As one of the most significant aspects of the structure of the underdevelopment of the host oil areas in the Niger Delta, rural poverty has thrown up some fundamental theoretical issues in the received literature in the field. The capitalist development promoted by Shell and other operating oil companies like Elf, Chevron, Agip, Pan Ocean and Mobil across the Niger Delta, further sheds light on the nature of usual poverty in these areas. Social infrastructures, like access road, which Shell provides in its host oil areas, might have helped open them for more economic and social activities. But that does not mean the host communities are not experiencing underdevelopment specially in terms of rural poverty and degradation of the environment. This is because underdevelopment does not mean a static condition. For, the oil companies cannot even further their capitalist accumulation in the Niger Delta area without first, putting in place, certain social infrastructure like access road. Shell's across roads are meant to connect its various oil/gas locations, but not necessarily to open up the areas for other examine activities. It is true that the people who live there take advantage of the road to move around. But that is by chance. In other words, there infrastructures only facilitate the exploitation of the oil producing areas, but do not really improve the material well-being of the people of these areas. The pertinent question therefore, is whether there is really any use of Shell's infrastructure to the people of its host communities who have no land to farm, nor stream to fish which are the result of its incessant pollution of their environment. In fact, it is in this context, that the nature of rural poverty of Shell's host areas in the Niger Delta is to be located.

That is not all. Land particularly rural land, which is regarded by the people of the oil areas and other rural areas of Nigeria as the root of their ancestral origin, is the most badly hit by environmental degradation. It has one of the most far-reaching consequences for rural poverty. Land provides the basis on which the major economic activities in the rural area including the oil areas are based. The politics of land therefore, provides an insight into the context and scale of rural poverty in the Niger Delta. For instance, the Mineral Act of 1914 as amended in 1967, and the Land Use Decree of 1978 all provided that, land where oil is explored, produced, transported and stored, is vested in the state. This means that what Shell does, is to apply for mining acreage for which it pays rent and other fees to the state. The same is true of other foreign oil companies operating in Nigeria. And should the activities of Shell degrade the environment, as is usually the case, it is only the state that is legally empowered to protest. Not even the indigenes of its host oil areas can protest since they are no less than squatters in their ancestral land by the provision of the state laws on land. And when they do protest, the state deploys its military to quell or crush them It does this on account of its materialism, and alliance with the oil companies.

Worst still, the state's sole ownership of land where crude oil is explored and produced makes Shell and other operating oil companies all the more reckless. To Shell for instance, there is nothing like an oil producing community in their scheme of operations. Instead, its entire oil bearing areas in the Niger Delta have come to represent a minefield. A mine-field, in the thinking of both Shell and the state, cannot be polluted. So, gas flaring, oil spillage, and dumping of untreated wastes on land/swamps are regarded as normal hazards expected in the process of finding and winning crude oil in spite of the fact that they pose dangers to the people. In essence, the worsening social and economic well-being of the people in Shell's host oil areas in the Niger Delta is facilitated by the state. Mobil, Chevron, Elf, Texaco, Pan-Ocean operate in the same manner. Locating rural poverty in the context of the contradictions engendered by state-Shell capitalism in the Niger Delta, therefore, poses challenges for the theses of W. Rodney[1] and F. Fanon[2] on the underdevelopment of Africa. That is, the state is no less guilty for Nigeria's Niger Delta underdevelopment than Shell. More details on this in the course of the discussion. Let us now present an overview of the economic geography of the Niger Delta as a background to a fuller understanding of the socio-economic and environmental consequences of Shell's mode of operation.

6.3. An Overview of the Economic Geography of the Oil-Bearing Areas in the Niger Delta

6.3.1. Geography[3]

As noted, the oil-producing areas in the Niger Delta are all located between latitude 4^{oN} - 6^{oN} of the Equator, and Longitude 4^{oE} -8^{oE} of the Greenwich Meridian. Generally low laying, the River Niger and its distributors drain the Niger Delta basin area into the Atlantic Ocean. The terrain is deltaic and has rich oil-bearing sedimentary rocks.

Two marked seasons, dry and rain, are known in the area. The dry season occurs between November and April, while April - October is rainy with a brief break of the rains in August. The Niger Delta has a characteristic tropical climate with an average annual rainfall of 200 cm in the relatively upland areas in Ughelli, Umuechem, Ahoada and Obigbo Communities, and 400 cm for the coastal communities of Nembe, Forlands and Bonny. Its temperature ranges between 34^{oc} and 45^{oc}.

The vegetation varies from the mangrove forest in the coastal areas, to the fresh water swamp and rain forest in the relatively less wet regions. The bulk of the activities of Shell and other oil transnationals is concentrated in this vegetation belt. The Niger Delta also has an average breadth of 15 kilometer of the continental shelf where oil exploration and production activities are beginning to take place.

Of Nigeria's total population estimated at about 120 million people as of 1997, 10 million, representing 8.3 percentage, are in the Niger Delta. To a very large extent, the geography of the area has shaped its economic activities.

6.3.2. Economic Activities[4]

Basically, the production system of the economy of the Niger Delta area is still subsistent in spite of the advent of the forces of transnational capitalism. It still subsists on agriculture with farming and fishing engaging about 80 percent of its total labour force. As the predominant occupation of the area, farming is carried out in most of the upland areas, with very little of it in the swampy areas. With slight variations, there are two marked farming seasons across the Niger Delta area which is largely dictated by the pattern of rainfall: November-May and July-December. The soil types are basically dry, sandy and humus, with the blend of these soil types varying from the coastal to the upland areas. Most of the crops grown during either of the farming seasons are, yams, cocoyam, pepper, groundnut, rice, vegetables and cassava. Rubber, the major export crop was introduced into the area by the British during the colonial period, and it is largely grown in the upland areas.

Fishing, next main occupation of the people in the oil-bearing areas of the Niger Delta after farming is commonly practiced in the coastal areas. The method of fishing is still largely traditional among the people; that is, using net and traps as opposed to trawlers. The swampy and part of the rainforests where most of the hunting is done, have large population of bush rats, grass cutters, pocuppines, antelope, monkeys, snails and reptiles. Elephants are also found around the Okpai and Umuoru Communities of Delta State where Agip is the dominant operating company. Salt-making is another economic activity of the people with most of its labour supplied by the family. Nature salt is produced from the aerial roots of the salty mangrove trees. In spite of the efforts of British rule to destroy the socializing basis of life of the people, their belief system, traditional ways of life refuse to die.

Generally, these geographical, economic activities complement each other in the development process of the area. I have followed conventions so far. Let us now analyse the impact of oil production on all this.

6.4. An Analysis of the Environmental, Social and Economic Consequences of Shell's Operation in Its Rural Host Communities

To be able to analyse the impact of Shell's activities on the environment economy and social-cultural life of its host communities in the Niger Delta, it is important the various stages involved in the exploration and production of crude oil are first understood. For, it is in these stages that a greater part of the consequences are to be found. Five stages can, at least, be identified right from exploration stage to when the crude oil is piped to the terminal ready for export. They are exploration, drilling, gas flaring, transportation, and storage. Let us begin the discussion with exploration being the initial step, noting its impact on the material well-being of the people of the oil areas.

6.4.1. Exploration

The degradation of the environment begins with the exploratory activities of Shell; and its contractor OSCs since a lot of the chemicals used during seismic shots and preliminary drilling are not only harmful, but the untreated wastes generated therefrom, are all dumped on land and into swamps. Worst still, Shell, its OSCs and their seismic crew often numbering 2000 in a single operation in an area, violate the cultural norms of the host oil communities in the process of searching for crude oil. All this has significantly disrupted both the economic activities of, the sociological basis of livelihood of the people who live in company's host oil areas. Some technical illustration is important at this point in order to bring into bolder relief, the impact of oil exploration on the environment and economy of the Niger Delta.

In a typical single seismic operation carried out by Shell particularly in an on-shore location for instance, about 15 hectares of land most of which are farmland, are usually delineated for prospecting. Thereafter, tramps (cables) are laid to link up the various points at which explosives were sunk into the soil. Finally, the explosives are detonated to enable the sensors which have been driven into the earth coast at internals, determine oil-bearing synclines. It is on these synclines that preliminary drilling is further net Water-based chemicals like EP-20, soda ash and pipe lax are usually used for drilling at this stage. Harmful as they are, the waters generate are not treated before dumping them on the environment. All this portends grave consequences for both the environment and the economy of the host community and its people.

For the past 30 years, there has been general complaints to Shell and the NNPC by the inhabitants of the oil-bearing areas of Ughelli, Isoko, Ethiope in Delta State, and those of Umuechem in Rivers State that, the sound generated from detonated explosives during exploration, has caused earth tremor, disrupted the natural process of soil sedimentation, caused the flight of animals from the area, and violated some of the traditional rites observed from the deities of the people. They also raised alarm over the pollution of their environment by the wastes. Yet, both Shell and NNPC representing the state in oil sector, made efforts to devise other means of finding oil. These are various levels of destruction done to the host oil areas regardless of whether it is a 'dry drill', meaning no oil is found, or not. In fact, the scale of damage is more severe when crude oil is found since the process has to be repeated all over. One of the major problems arising from oil exploration is land pollution, which, in turn, has heightened land scarcity in the host area where farming is the major occupation of the people. In the Isoko oil area where Shell accounts for 100 percent of the oil wells, its previous exploration activities has claimed over 15,000 hectares of farmland. The same is true of the Umuechem area of Rivers State where Shell had polluted about 10,000 hectares of farmland in earlier exploration work. By comparison, Elf is no less guilty in Ogbaland where its past exploration activities resulted in the pollution of 5,000 hectares of farmland. Nor is Agip any better than Shell and Elf on its land were polluted.

On the cultural environment of the area, the effect of exploration was equally damaging. The sociological basis of the culture of the people was usually violated with impunity, particularly by the workers most of whom were of different socio-cultural background. For instance, most of the oil workers willingly pollute the environment while poaching the resources of the area as evidenced in the use of dangerous chemicals like gamalin 20 to catch fish, and indiscriminate manner in which they killed totemic animals as delicacies. Table 6.1. summaries some of the totemic animals of the oil area that were killed by Shell workers.

The reverence of these animals/fish as shown above, stemmed from the traditional belief of the people that, they had played and still play very significant role in their cultural life and existence. Killing any of them at all by anybody much less by oil workers who are foreigner in the land, is believed to be capable of spelling doom for the people and the entire community. Hence, the poor harvest, prolong dry season, outbreak of strange diseases among

Table 6.1. Selected Revered Animals of Shell's Host Oil-Producing Areas of Bayelsa, Delta and Rivers States killed by Shell workers

Community/State	Revered Animal	Traditional Significance
Usiefurun (Delta State	Python	Repels marine demons against the community and its people
Bonny (Rivers State)	Python	ditto
Jeremi (Delta State)	Snail	Helped them to win tribal wars in the past; and still does
Ogbia/Nembe (Bayelsa State)	Python, fox and shark	ditto; and protect the people against evil forces
Evwreni (Delta State)	Crocodile and monkey	ditto

Source: Author's field works between 1989 and 1996.

other problems, that the people who live in the oil areas of Ogbia/ Nembe, Ughelli and Bonny have been experiencing in the last 40 years when Shell begun operation there, were believed to have been caused in part by the violation of the taboos observed for the various gods and deities. The majority of the people living in host oil areas fervently believe that, the protracted hardship and poverty in the land can hardly be reversed if the spiritual angle to it is not tackled. And the prospect for this is oil since both Shell and the state do not believe in that. What is more, no sooner had a rite to appease the gods been performed than more cases of violation recorded by the company.[5] The reckless manner in which Shell has been producing crude oil further compounds the degradation of the environment.

6.4.2. Production

Broadly speaking, drilling, flaring of gas, and transportation, make up the major stages involved in the production of crude oil. At each stage, the environment is polluted. So, too, is the socio-economic activity of the people particularly badly destroyed. Let us now turn to each of these stages in order to gain a deeper understanding of the depth of environmental pollution and rural poverty in Shell's host oil-communities.

6.4.2.1. Drilling

Following any successful exploration, is drilling. Be it on- or off-shore location, drilling definitely involves steps like the cutting of well formations and cushioning them, use of water-based chemicals most of which are harmful, and the management of the wastes generated from the process. Some of the water-based drilling chemicals like pipe lax, bentonite and EP-20 have been known to be harmful. The same is true of the wastes generated therefrom, which do not only degrade the environment, but also kill human beings. See Table 6.2.

In Delta State for instance, the inhabitants of the oil-producing areas of Afiesere, Ufuoma and Omavowve axis in Ughelli areas; Uzere, Olomoro, Owhe and Ogini communities in Isokoland all of which are operated by Shell; are like those in Agip's host communities of Otagba-Uno, Umuoru and Okpai in Ndokwa area, and those of Pan Ocean's host communities like Ogharefe, complained that the reckless manner in which crude oil is drilled has left their environment and economic activities severely destroyed. They summed up some of the observed consequences of pollution caused by harmful drilling wastes as follows:

The swampy areas are particularly easily and badly polluted by the harmful drilling wastes dumped by the oil companies. Save the crocodiles, alligator among other reptiles that can crawl to land, the rest die from intoxication. Other members of the faunal family are also killed by pollution. The danger of polluting the swamp is that, the wastes can easily mix with water and kill organisms as the toxin spreads fast to other areas. At times, it can spread so fast that communities 40 kilometers away from the point of initial pollution are affected within 24 hours. On land, the vegetation is particularly destroyed. In fact, the process of pollution continues as the wastes immediately spread to unaffected areas after rainfall. Worst still, the wastes percolate into the soil, thereby prolonging soil infertility and land crisis across the area. Pollution has destroyed our means of livelihood.[6]

To shed more light on this, case studies on selected host oil-producing communities that have been particularly badly hit are undertaken in chapters Seven and Eight.

6.4.2.2 Gas Flaring

In its natural state, crude oil carries a lot of gangue materials like natural gas which is associated with it; or land that co-exists with it. According to oil exploration and production engineers who the author spoke to at Warri and Port Harcourt in 1990, the associated gas would have to be flared in the process of extracting

Table 6.2 Selected Major Drilling Chemicals used by Shell in the Niger Delta and Their Consequences, 1996

Name of Chemical	Mode of disposal of Wastes	Consequences on the Environment	Remark
Persene drilliad	mixed with water before its usage; wastes generated were untreated and emptied into the environment	causes shortness in breath, and harmful to the fauna community, resulted in the out-migration of animals like gorilla and monkeys	prevalent cases of asthma; depletion of the game reserves
Soda ash	water-based; wastes untreated and dumped into the environment; unless clashed, the effects of the harmful wastes could remain for two decades	also caused difficulty in breathing for both human beings and animals	prevalence of asthma and pneumonia; depletion of animals like gorilla
Pipe lax	water-based; harmful wastes dumped into the creeks and on land; effect could last for a minimum of 25 years if not cleansed	pollutes the flore and fauna of the area; left fruits and vegetable polluted; both surface and sub-surface waters polluted; causes severe abdominal disorder if polluted fruits/ vegetable are eaten and the water drunk; and cancer of the skin if the polluted water is bathed with	several reported cases of abdominal ulcer, scabies, and skin cancer
EP-20	water-based; harmful wastes so, untreated and dumped into environment	inhaled odour causes instant death; too, if polluted water is drunk by both human beings and animals results in death; caused bronchial causes bronchial problems	prevalent cases of bronchitis; incidences of dead animals found in the area

Source: Author's field work, April 1996.

the crude oil. This partly explains why Shell, and other oil companies like Agip, Elf and Texaco flare gas in their oil locations. However, this does not mean that the flaring of gas will have to continue in the process of crude oil production as Shell does. It is for this reason that Shell and other oil companies operating in their home countries in Europe and America hardly flare gas soon after the production of crude oil starts. Which explains that Shell is still flaring gas in both day and night for the past 40 years in its host communities across the Niger Delta because the company receives the backing of the Nigerian state. Suffice it to note that, the flaring of gas is political and not that it cannot be stopped if the state so desires. More explanation is important.

Given the provision of the joint venture agreement, Shell is technically seen as a 'junior' partner to the state (NNPC), though it is the company that operates the joint venture which produces crude oil. Technically speaking, therefore, it is NNPC/Shell joint venture agreement that flares gas. That explains Shell's insistence on the need for the joint venture to hire other oil/gas companies with specialty on such capital-intensive projects like gas re-injection if it has to stop the flaring of gas. Furthermore, Shell insists that the cost of re-injecting gas will have to be borned by both the NNPC and itself, being partners to the joint venture in which the former has a higher participatory ratio. In terms of cost, it would mean that the state has to bear about 75 percent of it with the remainder paid by Shell. Shell is not alone in taking this path of argument in an attempt to pass the buck to the state. For, Agip, Texaco, Mobil, Pan Ocean, Chevron and Elf all do the same. However, while companies like Agip was about the first company to install gas plant at the Okpai field in Delta state, to help minimize the flaring of gas there, Shell has not considered doing so, contending instead, that it is all plot by other foreign oil companies with smaller concessions to break its leading position in the country's upstream oil industry. Shell's position on this, is mistake. Rather, the company feels it is too big to be pushed around by the state.

To the Nigerian state, the contention of Shell and other oil companies is not acceptable. But the question is: Can the state do otherwise? Caught up by the contradictions engendered by its mode of surplus extraction in the oil sector, the state can hardly enforce its Gas Re-injection Decree of 1979. For, as a landlord, it does not understand why the oil companies including Shell have to dictate the rule of operations. Rather, it decided to impose fines on any operating oil company that violates the Gas Re-Injection Decree. By 1998, it was increased from 50 kobo to N10.00 per 1000 cubit feet of gas flared. Shell was not deterred. In July 1999, it was raised to N20.00 per 1000 cubit feet of gas flared. Still, Shell was not bothered. Why?

Re-injection decree and the penalty have rebound effect on the state. Of the N20.00 fine paid per 1000 cubic feet of gas flared for instance, Shell would pay 20 percent, being its ratio in the joint venture, while 80 percent is borne by the state. In terms of the current ratio, Shell would pay 30%, Elf 5%; Agip 5%; NNPC 60%. On paper, therefore, it is still the state that bears the lion share of the cost. In practice, however, there is no official sharing that the NNPC ever paid fine for gas flaring. Officials of Shell claimed that it paid fine, but that it was expanded as production cost to the joint venture, meaning the state unknowingly pays for the recklessness of Shell. In fact, this is what other oil companies operating joint venture with the state do.

Across its host oil areas in the niger Delta, therefore, Shell has been flaring gas day and night for over four decades. Chevron, Pan Ocean, Elf and Agip are no less guilty. Of about 100 million cubic meters of gas flared by the oil companies in the Niger Delta annually, Shell accounts for about 70 percent. Of Shell's share, 80 percent of it is flared in Bayelsa, Delta and Rivers States, and the remainder in the company's other host states like Akwa Ibom and Imo. This means that, the oil communities of three states, namely, Bayelsa, Delta and Rivers, are worst hit. For instance, within Port Harcourt, Shell still flares gas both day and night for the past 30 years in Obigbo. Agip's flare sites located along Rumuola road in Port Harcourt were less than 500 meters from the homes of the residents of the area. This negates the rule that flare sites should not be cited in residential areas. The case in Ogoniland where Shell is the main operating company has been so well documented by MOSOP, and studies by foreign NGOs that it needs no further elaboration here. Bonny community is about one of Shell's off-shore location where the residential areas are virtually littered with flare sites. The question is: What is the consequences of all this for the material well-being of the people and their environment?

The impact of gas flaring on the economy and environment of Shell's host oil-bearing areas is grave. In fact, its magnitude can be better appreciated if gas flaring and its debilitating effects are analysed in conjunction with the impact of the continuous dumping of harmful drilling wastes on land and into swamps. For, as the oil companies dump the dangerous drilling wastes on the environment, killing plants and animals of the area in the process, the heat generated from the gas flare, simultaneously desiccates the soil and turns the hitherto rain and mangrove forests into derived savannah. This is the trend of environmental degradation across the Niger Delta with Shell's host oil communities particularly badly hit. Let us illustrate with one of Shell's host oil community in Delta State.

The Kokori Orogun oil and gas field of Shell has 30 wells. In 1970, Shell had four flare sites in the field, with one located at Erhoike (Kokori), and the remaining three at Erhabaro (Orogun). As the oil wells aged, they were reduced to two, with one each at Erhoike and Erhabaro as of 1997. Like in other Shell's on-shore host oil areas in the Delta region, while the well heads were on land, the flare sites were located in the swampy areas. All has come to further degrade the environment. Within the company's oil and gas field in the area, farming was prohibited in a predominantly farming community. Over 3,500 hectares of farmland and 600 rubber plantations covering about 2,500 hectares of swampy land and owned by 400 croppers were lost to Shell's gas flaring as of 1980. As the company's recklessness continued unchecked, additional 3,000 hectares and land were lost bringing the total hectares of land lost to 8,500 as of 1996. In all, over 2,000 families whose major sources of income were farming and tapping rubber were displaced. Most of the rivulets and creeks of the Okpare river criss-crossing the entire swampy areas and the traditional fish ponds of the people are located, were severely polluted. Some of them have either been dried up or lost their natural capacity to retain most of the members of the flora family; Palm trees, virtually all of which are wild, are particularly badly hit by gas flaring. Not only did most of them located within a radius of one kilometre from the flare point die, the few outside of this range that are still found at the time the author visited the area as of 1997; could no longer bear fruits due to the regular and intense heat from the flare. This has resulted in the drastic reduction in the production of palm oil in the area. In essence, two other means of livelihood of the people, that is, tapping rubber and palm trees, were severely destroyed.[7]

That is not all. Land, particularly arable farmland, was so polluted that it no longer support the growth of most of the crops like cassava, yam and cocoyam. Worst still, there were no marginal land. As an interim measure of survival, the people of the area have turned their compounds into farmlands. Even then, the yields could only sustain a family of 10 for one week at most. Bush rat, grass cutter and antelope that used to be common games in the area, had fled; meaning little or no animal protein was taken by the people. Nor were fishes caught as the streams were virtually dead biologically. One of the hallmarks of the displacement of the people from their means of livelihood, is that women, who still constitute the bulk of the local labour force in these areas, were compelled to sell their unskilled labour power as concrete carries at N50 per day for an average of 7 days in a month to bricklayers/ builders who showed up in the area whenever its indigene resident in urban centres, came home to build houses. But this was very irregular due in part to the crisis in the country's economy. On the other hand, most of the men, with the youths dominating, had to migrate to the cities in search of means of living.[8] Shell's mode of laying of oil pipelines has also destroyed most of the economic activities of the people in its host communities.

6.4.2.3. Transportation

Three main maps can be profoundly drawn in all oil-producing communities in the Niger Delta inclusive of those hosting Shell. They are basically, those of the state, local government, and the oil companies acreage and their flow lines. The third category of map delineates the boundaries of oil concessions and flowlines of the oil companies. Crude oil might not necessarily be useful at the spot where it is produced. As a result, oil pipelines are laid to convey the crude oil to places where it is further processed before it can be consumed. For instance, Shell has a network of oil pipelines stretching over 7,000 kilometers throughout all its location in the Niger Delta. By the provision of the Oil Pipeline Act of 1958 as amended in 1963, the oil companies, as explained earlier, obtain permit from the state for a right-of-way on which their oil pipes conveying both crude oil and gas are laid. Because the state is basically concerned with the revenue it can use the Act to generate, it did not really matter whether Shell for instance, had its pipes laid across farmlands, fish ponds, homes, and sacred sites. This is all the moreso since the state owns all land where oil is explored, produced, transported and stored. Thus, the state granted the foreign oil companies including Shell, a lease period spanning 99 years, a period during which no social and economic activity is permitted within 500 meters on either side of the way. Given the long period of lease, the thinking of Shell is that, land on which its oil and gas pipes are laid, virtually belongs to the company. Not even the reduction of the lease period to 30 years later after the amendment of the Act, changed the modality of Shell's operation on the parcel of land on which it has a right of way. All this explains why Shell's oil and gas pipes are not only laid bare on the ground and across residential compounds in Port harcourt, Umuechem, Bomu, Bonny and Nembe among other locations of the company in the niger Delta, but are not maintained. At Ogharefe where Pan

Ocean operates, it is not different from Shell's. Other companies like Chevron, Texaco, Agip and Elf are equally guilty of this.

The ecological consequences arising from the manner in which Shell's oil and gas pipes were laid are many and grave. As high pressured pipes laid bare on the ground, there have been cases of most of them bursting due to corrosion and lack of cathodic maintenance. Though farming is not allowed on the parcel of land within the right of any of the oil companies, inhabitants of the area ignore the warning and till such land most of their farmlands were polluted. In fact, the economic activities of non-oil producing communities that fall on Shell's oil and gas pipelines have not been spared as well. Let us illustrate on this with the Aleibiri community under the Sagbama Council Area of Bayelsa State.

Located along River Bomadi, Aleibiri is a predominantly fishing community. Thought crude oil is yet to be discovered in the area at the time this study was conducted, Shell's Beneseide/ Forcados trunkline, and the accompanying complex network of its oil and gas pipelines criss-crossing the Aleibiri creeks, have profoundly impaired fishing activity there. Westminster Dredging, contractor company to Shell, that dredged the Ayaoba Cana located west of Aleibiri town, has left it too deep for the people to fish. Worst still, the major traditional fish ponds and lakes namely Borudonu, Opudonu, Aglab and Lalpu located within it, were all destroyed following the dredging of the canal. As a consequence, the natural course of migration of fish from the Atlantic Ocean was redirected away from the area. This has resulted in poor catch by the fishermen. As a major economic activity of the people, fishing has virtually stopped in the dredged canal not only because it became too deep for the people to fish, but the oil and gas pipes laid underneath the water compel the state/Shell to stop further fishing in it. Not that. Its depth of about 500 meters made it unsafe for hand-dug boats used to ferry people and for trading, to ply. All this has impacted negatively on the ability of the inhabitants to eke out a living.[9]

Shell's tank farms and manifolds have also been found to have disrupted the economic activities of its host communities, particularly those hosting its terminals. For instance, Shell built its terminal at Bonny on 1,500 hectares of dry and land belonging to the Jumbo Family. The company had the option of reclaiming land given the fact that dry land is scarce in the area. Even the state ought to have compelled Shell to reclaim land for such project if it were interested in the sustainability of the environment and economy of the oil-bearing communities in the Niger Delta. Taking such dry land, resulted in the displacement of members of the family who subsisted on it. Though the company paid compensation, it was a pittance. As the author's investigation revealed, the compensation money was misappropriated.[10] Furthermore, all trunklines in Shell's East division, terminate at Bonny terminal. In and around Bonny too, Shell has numerous storage tanks where crude oil is stored before it is exported. Shell's extensive network of trunklines, tank farms and manifold in and around Bonny had impaired fishing and other economic activities of the people of the area. Constant emission of effluent from the tanks and houses during lead have polluted the water thereby compounding the problem of survival of the people. Shell's storage tanks at its second terminal at Forcados, Delta State, with a capacity of 500,000 barrels of crude oil, had its underneath value loosened and the entire content was emptied into its immediate environment in 1979. Apart from the widespread pollution of the immediate and neighbouring communities, an estimated value of the fish killed in that singular incident, was put at N2 million. Being the most severe spill ever recorded in the area, its impact on fishing was still serious at the time of the study. How oil spillage impacts on the environment and the socio-economic activities of the people is now examined.

6.4.2.4. Oil Spillage

As noted, because both the Nigerian state and the operating oil companies regard the oil-bearing areas of the Niger Delta as a minefield, little or no attention is paid to oil spillage. That explains in part why Shell, the most guilty of all incidents of oil spillage across the Niger Delta area, has never really cared to clean up the oil mess in its host communities after any spillage. Nor has the company assessed the resultant effect of oil spillage on the social and economic lives of the people. The same is true of Agip, Elf and Texaco. All this explains why spillages in which between 10 and 10,000 barrels are lost are regarded as 'no spill'. It does not mean, however, that the loss of 1.5 million, and 2 million barrels spilled at Bomu (Shell) and Finiwa (Texaco/Chevron) in 1970 and 1980 respectively, and in a single spillage, really mattered to both the oil companies and the state. Instead, what was of concern to them, was the loss of revenue. Rather than undergo a rethink of their conception of the oil-producing areas, both the state and the oil companies still conceive of them as areas where oil is to be explored and produced only. Furthermore, oil spillage occurs at the same time other forms of pollution arising from exploration, drilling, gas flaring are taking place. Because it has the most debilitating effect, oil spillage appears to conclude the process of annihilation of the environment and economy of the host oil

Table 6.3 Observed Causes and Effects of Oil Spillage in Shell's Host Communities in Bayelsa, Delta and Rivers States, 1989-1995

Source: Author's field work, 1989-1995

NB: Bayelsa was part of Rivers States at the time of the study.

Major Causes of Oil Spillage in the Area	Consequences	
	Flora Community	Aquatic Organisms
Blowout, equipment failure, human error, engineering error, natural causes, sabotage, thirty party, accident and unknown.	Trees mortality defoliation, bark, fissuring, seedling, mortality, stunted leaves, deformed growth, chlorosis, heavily polluted banks of creeks and rivers and estuaries, killing of herbal seed weeds like water lettuce	Constant de-oxygenation of the water resulted in the asphyxiation of plank-sons: extensive pollution of the esturaries and the wetland caused the exinction of cat-fish, disappearance of sea-cow (manate), deformed eggs of fish and low hatching; crocodiles and alligators almost extinct; depletion of the mollusion family-snails, shellfish, periwinkles, crabs, and squids.

communities in the Niger Dealt, a process which began with the imposition of British rule in 1900. Which means that, the depth of rural poverty inflicted on the people living in the oil-bearing areas is such that, the people are left with little or no capacity to regenerate themselves socio-economically.

From the author's survey of Bayelsa, Delta and Rivers States, being the states that were worst hit by oil spillages, ten causes were identified to be responsible for oil spillage. They were: blowout, equipment failure, human error, engineering error, natural causes, sabotage, third party, sand cut, accident and unknown. Of these causes, however, equipment failure, particularly ruptured oil pipelines with most of them in Shell's locations, accounted for 80 percent of the reported cases.[10] This is because the ruptured oil pipelines were and still laid bare on the ground for the past 40 years without any maintenance except when they burst. See Table 6.3 for details.

What is more, 80 percent of the activities of the oil companies in the three major oil producing states, were and are still on-shore with Shell as the dominant company. As noted, Shell's first entry into the area and for a long period of monopoly until 1960 when other oil companies like Mobil, Chevron, Texaco were allowed to operate in the country, provided the company the chance to corner all rich oil-bearing on-shore locations across the Niger Delta. But that partly explains why Shell also accounted for 80 cent of an annual average of 3.5 million barrels of crude oil spilled through burst pipes, which corroborated an earlier study on oil spillage in the Delta between 1976-1980 in which Shell also recorded 598 out of the total 784 cases of oil spillage; and 900, 899 or 74 percent of the total 1.8 million barrels of crude oil spilled.[11]

Sabotage, is strictly speaking, a reaction of the people of the Niger Delta to the recklessness of the state and oil companies in their host communities. It is borne out of the feeling of the people, who bear the social and economic brunts of oil production. They contend that they will cause both the state and oil companies to lose crude oil even if they endanger their lives in the process. Being most notorious for oil spillages and gas flaring, Shell remains the worst hit by sabotage by the aggrieved inhabitants of its host oil communities across the Niger Delta. Ascertaining the accurate number of barrels of crude oil lost through sabotage in Shell's locations, is difficult since pipelines were usually cut at remote points beyond the surveillance of both the oil workers and the security officials of the state. Nevertheless, since 1975, Shell had reportedly lost a conservatively estimated average of 2.5 million barrels of crude oil annually due to sabotage, and all of which was lost to the environment.[12] Human error was also found to have led to the blowout of well heads. So, too, have engineering problems encountered in the platforms, well heads, and flow lines, caused millions of barrels of crude oil to spill. It was revealed by some of the field managers of the OSGs working for Shell in Nigeria that, incidents of blowout and sunk rigs stemmed largely from these errors. In 1980 for instance, Texaco and Chevron, and their oil platform at Finiwa, blow out with over 2 million barrels of crude oil all lost to the environment. In 1985, Shell had, for the first time, one of its off-shore rig in the Bonny oil field sunk, drowning its entire crew. Not only was the information about the incident suppressed by both the company and the state, the sunk rig was still not recovered at the time this study was done.

Since 1970, there was hardly any year when Shell did not cause crude oil to spill in its host communities across the Niger delta. More than other forms of oil pollution, oil spillage has the most devastating impact on the environment and economy of the oil areas. See Table 6.4. Summaries the consequences. It will be repetitive to document all the incidents of oil spillage. Rather, case studies of spillages in some of the oil-producing communities in Delta State will be used to shed light on its causes and impact.

Of all the on-shore locations of Shell in Delta State, the Isoko oil communities have been one of the worst hit by oil spillage. Based on the author's survey of Isokoland, about 600 cases of oil spillage were recorded by Shell between 1980 and 1993 with Uzere, Oweh, Olomoro, Ogini, Igbide and Irri particularly badly hit. Ruptured flowlines was identified as having caused about 95 percent of the spill. Though it was difficult having an accurate figure of the volume of crude oil spilled due in part to the ignorance of the people, and all the spilled crude percolated into the soil, about 15 million barrels of crude oil were estimated to have been lost by Shell in the area for the period under review. Also, its impact is so pervasive that all socio-economic activities of the people were badly affected. Fishing, which was the most important economic activity of the coastal oil communities in Isokoland, was almost extinct, mud fish and cat fish, two of the most common fishes caught in the area, had become extinct at the time the study was conducted. The crustaceans like water snails and periwinkles were significantly decimated too. So, too, has the population of crayfish/ lobsters plummeted. Crocodiles and alligators among other reptilian creatures for which the Isoko area is noted, have also fled since their habit is not only disrupted, but their prey and even themselves, are all being increasingly decimated by pollution.[13]

Table 6.4 Consequences of Crude Oil Pollution on Selected Flora and Fauna of Shell's Host Communities in Bayelsa, Delta and Rivers States, 1997

Flora and Fauna	Significance	Remarks
Cocoyam	Carbohydrate source, major staple food and was widely grown	declined productivity since 1975, and virtually getting disappeared since 1980
Electric fish	one of the dominant fishes caught in the wetland, and a major source of protein	became totally extinct since 1979
Manatee or sea cow	one of the rare fishes found across the Niger Delta but in the area major source of fat and protein; its fat used for cooking and medicinal purposes	decrease in its population was first noticed in 1975, and became extinct since 1980
Cat fish	one of the predominant fish caught throughout the wetland: a source of protein, and used for spiritual sacrifices	virtually became extinct since 1979
Hippopotamus	one of the rare sea mammals in the area; source of protein and fat; and spiritual purposes	migrated out of the area since 1978 to the Atlantic Ocean
Shark (fish)	It is totemic, and a major source of protection for the Ogbias in case of boat mishap or drowning in the river or sea	migrated from the area to the Atlantic Ocean

Sources: Author's field work, November 1997.

Pertaining to land, oil spillage has polluted quite a considerable hectares of farmland in the Isoko area. Farming, the most dominant economic activity of the people, has been impaired as more farmlands were lost yearly to oil spillage. As a minefield, Shell has never cleansed spilled crude oil, nor did it carry out post-spillage study to determine the extent of damage with the hope of restoring normalcy to the soiled and economic activities of the people. A family of six squatted on 100 x 50 meters of marginal farmland as opposed to 500 x 200 meters of farmland before Shell began operation there. Not only that yield from such marginal land was estimated at about 300 tubers of cassava, which if processed into garri, could only sustain a family of six for ten days. Yellow yam, a specie of yam generally grown in the area, could no longer be planted as a result of soil pollution. The same is true for cocoyam, and water yam, all of which are staple food of the people.[14]

All this, together with other forms of pollution, have implications for the income of the people most of whom are predominantly fishermen and farmers. The majority of them were unable to meet most of their social obligations to their families and wards. For instance, a lot of them could not pay school fees for their children. With farming and fishing almost grinding to a halt, the majority of the youths were displaced, forcing them to migrate to the cities in search of casual jobs which they hardly find. Even those who chose to remain at home, were not employed by both Shell and its OSCs. Nor did the state introduce schemes aimed at absorbing some of the jobless youths. Suffice it to note that, all has placed the youth at risk both from within their local economy and from outside of it; that is, Shell and the state. Because both Shell and the state were not really forthcoming with projects aimed at reversing the underdevelopment of the area, the people have decided to embark on self-help projects such as building town halls, secondary schools, community scholarships, roads and market halls.[15]

The nature of rural poverty and its trend in Isokoland as explained above, pervade the Okpai and Umuoru oil areas in Ndokwa Council Areas where Agip operates. The same is true of Shell's other host communities like Afremo and Saghara in the Bomadi Council Area; and Ajekota, Umolo, Okpare, Utorogu under Ughelli oil communities. The worsening incapacities of the people who live in these areas to alleviate their socio-economic hardship, deprivations, much more eradicating them, have triggered of chains of reactions. Reactions that have taken the form of protests, peaceful and violent. Let us now examine the reactions of the inhabitants of the oil-bearing areas.

6.5. Reactions of the Inhabitants of the Host Oil-Producing Areas

6.5.1 Clarifying the Context of Protests

To begin with, it is crucial to note that, not all protests in the Niger Delta are directly caused by the oil companies' mode of operation. The Nigerian State has, through its politics, been responsible for the agitation such as the violent conflict between the Ijaws and Itsekiris which arose from the relocation of the headquarters of Warri South Local Government Area from Ogbe-Ijaw in Warri to Ogidigbe in 1997 by then Abacha junta. Therefore, it is important to first indicate that only protests arising from the operation of Shell and the state are examined here. Thereafter, the context in which protests are organized in the oil areas of the Niger Delta is explained in order to have a clearer understanding of what they have set out to achieve especially if they are really aimed at halting environmental pollution and reversing the trend of poverty in these rural areas or not.

From all indications, what has been at the heart of the raging protests in the Niger Delta area, is no less than the contradictions engendered by state-transnational capitalism as manifested in the underdevelopment of the oil areas. This can be explained at two levels. First, is the crass ignorance among most who live in these areas, of the magnitude of the ecological crises caused by the operating oil companies. This has been compounded by the fact that, the majority of these people have little or no access to Shell and other oil companies, and the state, explaining in part why they could not lodge their grievances with the appropriate quarters. Even in cases when they had such access, both Shell and the state rebuffed them. At-the-risk-youths who constitute a strong revolutionary force often used to disrupt the exploration, production, transportation and storage of crude oil have themselves, proven equally relatively ignorant of both the nature of the state/oil companies alliance and the extent of damage done to their ecology and economy. These and other limitations of the people who bear the brunts of ecological crises arising from oil pollution, have dictated an aspect of the philosophical context of the protest. Limitations that have been taken advantage of by those I call 'conflict entrepreneurs', who, given their wide contracts with the state, Shell and other oil companies, have used protests to advance their selfish political and economic interests. Pretending to be on the same side of the struggle with the down-trodden masses of the host oil communities, they mobilize the youths to disrupt the operations of the oil companies ostensibly to get noticed by both the state and the oil companies. Although they decry Shell/state-induced poverty in the rural oil-rich areas, they do not really concern themselves with its alleviation. For instances, one Francis Okpozio, deputy speaker in the defunct Bendel State House of Assembly, 1979-1983, and a member of the late General Abacha's Constitutional Conference, was one indigene of the Isoko area who was reported to have mobilized youths to protest against Shell in the area, demanding for reparation from the company. He was also reported to have led the youths to seize seismic equipments of Shell's OSC's too. But, he abandoned the struggle soon after he was in government and at the conference. Okpozio was not alone, virtually all the traditional rulers and contractors to Shell,

OMPADEC and PTF are all part of the clique of conflict merchants across the Niger Delta. These conflict merchants, who guise as representatives of the oil communities, also lobby for compensation funds from Shell and other oil companies insisting quite often that cheques be even issued in their names. That is not all, they also get the oil companies to award contracts to them as gratis, promising to pacify the people so that oil operations will be hitch-free. Without an exception, the leaders and members of most of the associations/groups protesting in the oil areas, notably, the Association of Warri North and South Oil Producing Communities; Association of Contractors of Oil Minering Producing Communities; Traditional Rulers of Oil Mineral Producing Communities; Association of Oil-Producing Communities Leaders; and Nigerian Association of Mineral Producing Areas Development are opportunists who are concerned with what they can get, and not really interested in halting the degradation of the environment of the Niger Delta and improving the well-being of its people.

Second, is the issue of minority people, their rights and the politics of their underdevelopment. Minority politics in Nigeria dates back to the colonial period. The discovery of crude oil in the Niger Delta in 1956 did not really ignite minority politics per se. Rather, it is the deepening marginalization, deprivation and pillage of the oil-bearing areas in the Niger Delta, while oil proceeds are used to develop other non-oil producing areas in the rest part of the country that has re-kindled minority agitation in the area. The Northern part of Nigeria, especially Abuja, Minna, Kaduna which are non-oil producing areas as of 1999, are being developed from oil proceeds to the detriment of the actual oil areas in the Niger Delta. This has helped to renew minority agitation in the Niger Delta. Ken Saro-Wiwa of blessed memory, and other founding members of MOSOP, had cried against this social injustice meted out to the Niger Delta people by the central government. But proponents of this position are being severely criticized by a group of Nigerian scholars, most of whom I prefer to call experts of 'fiscal federalism' as represented by the late Ojetunji Aboyade, a renown economist and adviser to then Babangida junta. To Aboyade for instance, crude oil is God's gift, and therefore, a national wealth which should be used for the development of the Nigerian federation. In essence, there is really no community that can be called an oil area and therefore, it cannot be laid claim to. I think it is a mistaken position, for even in the period when both cocoa and groundnut once accounted for the country's major foreign earnings which Aboyade and his likes had cited to support their argument, the environment was not really degraded. Not even the production of solid minerals would be degradation of the environment of the host communities be compared with oil pollution. For instance, mining lakes could be refilled and later converted into a farmland. But scientific findings show that, soil polluted by crude oil would take no less than 25 years to recover. The incidences of frequent pollution from other sources such as gas flaring and dumping of harmful wastes on land, make the chances of the environment ever regenerating very remote. In fact, the thinking of this group of economist has largely influenced the state's response to the plights of the oil-producing areas in the Niger Delta. That partly explains why the inhabitants of these areas can be referred to as economic saboteurs, terrorists, and enemies of progress by the state managers. Hence, protests are crushed with military despatch. However, rather than suppress agitation in the oil areas, the use of the state military might have deepened protest there.

Final, is the buck-passing tactic deployed by both the state and the oil companies that actually degrade the environment. As stated earlier, by the provision of the joint venture agreement, neither Shell nor the state can legally be held liable for polluting the environment. Instead, it is the agreement that pollutes. But as the technical operator of the NNPC/Shell joint venture, it is the company that produces crude oil for the partners. In effect, it is Shell that pollutes the environment. This is the thinking of the state and blames Shell for the wreckage done to the ecology and economy of its host oil areas. Shell thinks otherwise, blaming the state for its inability to own up to its shortcoming. Shell also insists that it is the joint venture partners that should be held for the pillage of its host areas and not the company alone. However, while blames are being shifted between the state and Shell, the inhabitants of the area know that the real pollutant is Shell, particularly as the main operator on the ground. They disrupt the operations of Shell therefore, so that the state is provoked into action. But that has not helped matters as the action of the state has, on balance, sought accommodation with Shell, while unleashing anguish on the people.

It is in the context of the above analysis that protests in the oil areas of the Niger Delta inclusive of those serving Shell is to be located. Rather than being united in confronting Shell and the state, protests are fragmented, at collision course with each group, and revolving around the individuals and leaders. In fact, there is no single protesting group in an oil area representing the overall interests of the people. Every leader or group is out to protect its own selfish interests and those of their community. Due largely to the idiosyncracies of the leaders, poor coordination, lack of finance, lack of contact with the international environmental non-governmental organisations, protests against Shell and the state have not yielded expected results. In fact, both Shell and the state have taken undue advantage of the problems that the agitating groups are faced with, to break their ranks and solidarity. But one thing is so clear at the moment. That is, agitations in the Niger Delta have deepened, disrupting the whole oil sector of the country and its revenue base, and the global oil market. There have been age-long protests between the Ijaws, Isokos and Urhobos. This means that, not all protest is directed against the state and oil companies. There are cases of intra-community oil related conflicts too. But in all, three main patterns of protests are discernible, namely, legal, demonstration, and sabotage/hostage taking. They occur simultaneously in different oil communities and, at times, by solidarity, making its impact more felt by the state and Shell, among other oil companies.

Legal Actions: As law abiding citizens, the inhabitants of the oil-producing areas still believe that the law court is their last hope in settling the problem arising from the pollution of their environment. The initial reaction to the pollution of their environment through the reckless activities of Shell, was for the people to file legal suits against the oil company and the state. As noted earlier, the state's laws governing its petroleum industry seek to protect its mode of capitalism in that sector. Not only that, they equally shield Shell and other oil companies from litigations arising from land and public nuisance. In fact, the issues taken to court by the aggrieved inhabitants of the host oil areas, were and are still bothered on public nuisance, negligence and public interest, issues which are too technical for them to provide concrete evidence. It is not that evidences of negligence cannot be provided by these people. Rather, with the basis of their income almost completely shattered, they cannot hire, Botanists, Zoologists, Biochemists, Microbiologists among other scientifists who can carry out detailed research on the ecological crises caused by Shell's activities, and provide proofs to support their legal actions. What is more, the state has always invoked the provision of the Land Use Decree of 1978 to nullify most of the suits filed against Shell by the inhabitants of its host oil areas. Without a *locus standi*, it is difficult for a squatter on the state Land to file suits against Shell for polluting land where he/she lives.

In 1989, for instance, the Ivbrogbo people in Isoko Council Area of Delta State, sued Shell at the Oleh High Court for polluting their farmlands and traditional fish ponds, claiming N9 million as damage. The presiding judge invoked Section 1 of the

Land Use Decree to nullify their case.[16] Similarly, the Ugborikoko community of Warri South Council, had previously sued Shell in 1981, for flaring gas which polluted their land, water and places of abode, praying the court to cause the company to stop oil production there to save them from the health hazard; and to enable them reproduce themselves socio-economically. Again, it was struck out by the presiding judge, who ruled that since crude oil was and still the major source of income for the Nigerian government, he cannot stop Shell from flaring gas. He could not even request Shell to redirect its flared gas away from the homes of the people.[17]

Across its host oil-producing communities in the Niger Delta, no legal case filed against Shell for charges of environmental pollution has been worse t\at the time this study was conducted. This will remain so for years to come as the state backs up shell, and other oil companies. It is not necessary, therefore, to document failed legal cases in other areas. That is not to say that it is an all-round victory for Shell, other oil companies and state since the people have resorted to other forms of protests such as demonstration, while legal actions are still taken as the necessary first step.

Demonstration:

Protests across the host oil bearing areas in the Niger Delta have taken two dominant forms, namely, peaceful and violent. With the exception of the uprising by the MOSOP in Ogoniland since 1993 which culminated in the execution of Ken Saro-Wiwa and eight other Ogonimen in November 10, 1995 by the late General Sani Abacha junta, and the youths agitation in Isokoland, most of the demonstrations in the Delta region has been relatively peaceful. But since the hanging of nine Ogoni environmental actvitists in November 1995, there has been simultaneous eruption of violent protests in the Niger Delta with the majority of them in Shell's host oil communities in Bayelsa Delta and Rivers States. Let us first explain peaceful protest in the Niger Delta area, paying particular attention to Shell's host communities.

Realizing that the court is part and parcel of the state structure, which explains why previous legal suits filed against Shell have been struck by the presiding judges, the inhabitants of the oil-producing communities, decided to embark on peaceful protests. Their thinking is that, it will provide them a forum to discuss issues of neglect and pollution directly with the officials of both Shell and the state. It begins with the affected communities sending representations to Shell, stating their grievances and demands, and suggesting a joint meeting with the company during which issues of pollution and marginalization are expected to be discussed. All is done with the aim of creating an enabling environment in the community which will permit Shell to operate. The Umuechem Community, MOSOP, Isoko Youth Movement, Ijaw Youth Movement, among other protesting groups in the Niger Delta, have all tried this method, but all was to no avail.

As negotiation seemed not to be yielding any positive result, direct confrontation was considered the next option. Still non-violent, confrontation has usually taken the form of barricading access roads leading to Shell's oil and gas installation. Armed with leaves, the youth and women lay siege on the company's access roads, chanting songs reminiscent of the agony its operations have caused them; and as well, calling for an urgent stoppage of the company's reckless mode of operation. The author's checks have shown that, at this point, Shell would not only grant the agitated people of its host areas audience, but also pacify them with promises of providing social amenities like pipe-borne water, electricity and tarred roads other than access roads. Records have shown that, Shell's intention is to douse tension so that it can operate without fear of reprisal from the people , but fail to fulfill any of its earlier promise. Shell has acquired the notoriety of reneging on its promises to the people of Umuechem, Nembe, Isoko, Ogoni, Urhobo, Ijaw to even threatening to invite the state to deploy its military to quell any manner of protest that would disrupt its activities should the people refuse to dismantle their barricades. Far more than a threat, Shell did invite the state, reminding it that the company was only a junior partner in the joint venture; and that should the state fail to act promptly, much revenue would be lost as a result of such barricades. Because the state is basically concerned with how much revenue it can generate from oil production and export, it usually sided with Shell by deploying its military to either nip in the bud, or crush and form of protest against the company. Suffice it to note that if the law court that ought to be the last hope of the people was no longer reliable, and peaceful protest seemed not to be yielding positive result, the only option was to violently confront the company and the state. It is more or less the last resort, undermining the consequences of it on the lives of the people, their economy and environment.

Violent protests have been in various forms. First, is the category that is typified by hit and run tactics. Because most of the individuals and groups involved in this type of attack lack resources and connections to sustain prolonged protest, they usually mobilize few youths most of whom were drop-outs from secondary schools and thugs from the rural communities, to barricade Shell's access roads and its oil and gas field, disrupting oil exploration and production, and halting other vehicles from plying the roads. This was usually done for about 21 days. All was to cause panic in the area. Once this is achieved, Shell would not only suspend operation there, but would be forced to negotiate with the leaders of the groups involved. At times, Shell would settle them with money so that it resume operation there. Before 1993, this was the trend of protest across the Niger Delta. But in most cases, the company would invite the state, which in turn, would deploy its military. This has given rise to a more violent form of protest.

Next, is prolonged violent demonstration. Unlike the hit and run tactics, demonstration of this type is usually carefully planned in order to have great impact particularly in reducing the capacity of both Shell and the state to produce oil and wreck further havoc on the environment and economy of the host areas. Individuals and leaders of groups who are involved in protest of this scale, try to ensure that fifth columnists and opportunists are not allowed to hijack the struggle for their selfish gains. It is relatively better organised than the first category in the sense that its leaders appear to have better connection within and outside of the country; and therefore, able to source for funds from international non-government organisations. Because of the relatively better exposure of its leaders, the youths, who constitute the fighting force of such protests, are well informed of the purpose of the struggle, better organized and focused. On average, it is only the MOSOP that fits into this category. This is not to say that, personal idiosyncracies of the leaders did not constitute a problem for the MOSOP. Intra-MOSOP crisis for instances, led to the splitting of the Movement into factions, the most militant faction of it was led by the late Ken Saro-Wiwa. Notwithstanding these problems, MOSOP was able to cause Shell to suspend operations in Ogoniland since 1993. Even after the hanging of Ken Saro-Wiwa and other Ogoni environmental activists in 1995, Shell was yet to resume operation there at the time the study was conducted. The activities of MOSOP have been so fully documented both locally and internationally, that they need not detain us.

In the West of Niger Delta, the Isoko oil communities have been noted for protracted protests, though not on the scale of the MOSOP. Not really an equivalent of the MOSOP, but has equally developed the capacity to sustain the struggle, is the Isoko Youth Movement (IYM). The IYM provides the umbrella for all youths in the various clans who are poised to confront Shell. In 1994, the youths in Uzere exiled their traditional ruler for receiving gratis from Shell, and sacked the leaders of the Uzere Youth Movement for charges of corruption and non-commitment to the struggle. Like the MOSOP in Ogoniland, the IYM wanted the exit of Shell from the Isoko oil area for polluting their environment. Agip which operate at Irri, has not been spared by the operators too. Though its operation in Irri area is far smaller than Shell's, the company is still recalcitrant and continues to flare gas there. Organizationally, the youths in the various oil communities in the Isokoland have close touch with each other so much so that sympathy demonstrations were easily carried out once it started in one clan. Given its prolong conflict with Shell since 1975, it might be the second group that will force Shell out of its area in Delta State, the first being at Oben in 1983. The Oben case was short-lived since the concern of people was to stop Shell from further oil production after one of its well-head blew-out killing three workers, and coupled with its track record of recklessness in the pact. The IYM has been agitating for the pullout of Shell from Isokoland. In this process, the Movement has gone to the extent of dismantling all the stooges of Shell in the area.

Third, and final, is sabotage and the taking of oil workers knowledge. Both methods of protest have impacted negatively on the capacity of Shell to continue with oil production much more exploring new wells. This has reduced the daily output of crude oil produced by the company, leading to a cut in the revenue accruing from oil export to both the Nigerian state and Shell. Let us elaborate more on each of them to bring into bolder relief, how effective they have been, beginning with sabotage.

As a mode of protest, sabotage has remained most dreaded both by Shell and other operating foreign oil companies, and the Nigerian state. Resorted to as the last line of action, the youths in particular and other fighting force of the various movements in the host oil-bearing areas generally felt that it was only through this means that Shell can be brought to the negotiating table and forced to meet with their demands. Dating back to 1975, the sabotage of Shell's efforts to find and produce crude oil is a closely knit network of oil workers some of whom are in the employment of Shell, and others, members of the agitating communities. It requires the services and collaboration of some of the oil engineers for sabotage to succeed. The fact that the processes involved in oil production are so specialized that no mistake is allowed makes sabotage a specialty. For, any error will not only expose the collaborators from within, but might have the saboteurs consumed in the ensuing gas/oil spill. For instance, before an oil pipeline is cut, a long period of surveillance is required to ensure that the point at which it will be ripped open is far from the view of the security personnel of both Shell and the state. It is even more important to ensure that the values controlling the well head and the pipelines from the various wells to the flow station, are all turned off. The author's check revealed that, it was during the period when the pipelines, well heads and flow stations were under repair that the pipelines were cut. This information was usually conveyed to the agitators by the oil workers who connive with them. By the time the normal pumping of oil is resumed, there will be spillage at the various points where the pipe is cut. At times, it will take Shell about seven days or more, depending on the terrain to discover the point of leakage. Several oil pipes were usually cut in one operation, forcing Shell to shut down to stop the spillage in most cases since it will be difficult to locate them almost immediately. As noted earlier, it is not that the people do not understand the implication of cutting oil pipes for the environment, particularly their economy and health. They do. Rather, it is an action borne out of past protracted but futile peaceful efforts that they have made to stop Shell from further polluting their environment. In essence, the people decided to cause oil to spill, believing that is one of the ways through which Shell and the state will loose revenue. Even the Anti-Sabotage Decree of 1975 that prescribed death penalty for anyone caught cutting oil pipe did not scare them.

As explained earlier, cutting of oil pipes dates back to the mid 1970s. With the launching of the Ogoni Bill of Right in 1991, and subsequent hanging of nine Ogoni environmental activists in 1995, this form of sabotage became prevalent. From oil spillages caused by sabotage for instance, there was a drastic fall from a total daily output of 2.2 million barrels of crude oil to 900,000 barrels in 1991. And between 1991 and 1995, sabotage-induced spillage so deepened that daily production could not get to one million barrels. As a consequence, the country's annual total crude oil production dropped from 800 million barrels to about 300 million barrels within the period under review. Since 1995, it has been an all fall due not only to sabotage, but to the communal disturbances mostly in Bayelsa, Delta and Rivers States. Since Shell is the predominant operating company in these areas, Shell was particularly worst hit. By April 1994, for instance, sabotage has caused Shell to loose 100,000 and 50,000 barrels of crude oil daily in the Western and Eastern divisions respectively, excluding Ogoni land where it has been losing 28,000 barrels per day since 1993. On average, the company had lost 158,000 barrels per day between January and April alone in 1994, and lost more after the hanging of the nine Ogonimen. From Shell's total estimated 1,400 oil wells spread across the Niger Delta, 50 percent of them were closed down between 1991 and 1995. There were no signs that Shell would resume operations in the affected areas particularly in Ogoniland as tension was still high not only in Ogoni community but in others that have embarked on sympathy protest following the hanging of environmental activists on November 11, 1995. Other operating oil companies, notably, Chevron, Agip, Texaco, Pan Ocean and Elf have also had about 20 percent of their oil wells shut down due to communal disturbances spilling over from Shell's host oil communities. For instance, about 20% of Chevron's 250 oil wells were shut down; and the same percentage for Agip's 125; Elf's 75; Texaco's 40 and Pan Ocean's 10 oil wells.[18]

Apart from the cutting of oil pipelines, other forms of sabotage adopted by the people to frustrate Shell's operations included, deliberate industrial accident, dumping of heavy oil tools into the sea, deliberate damage to machines, and storage tanks. All these acts of sabotage were carried out by a small squad of the various agitating groups some of whom are also workers of Shell or its servicing companies. Since the brief of this group of agitators is to cause Shell to suffer huge loss, either in the form of spilled crude oil or loss of expensive oil tools, all devices are employed to succeed. Based on the author's checks, there were cases in the past, when Shell had some of its equipment hauled to install platforms at off-shore location, worth millions of United States dollars, dumped into the high sea. The underneath values of most of the company's storage tanks at the terminals have been one of the prime targets of the agitators. The loosening of the underneath value of one of the Shell's storage tanks at the Forcados terminal in 1979 was one operation Shell has found it pretty difficult to fish out those responsible for the act because it is a conspiracy in which some of the company's staff are involved. If the various modes of sabotage have made the Niger Delta gradually becoming unsafe for Shell in particular, and other oil companies generally to operate, then the taking of oil workers hostage, is forcing some of them to maintain skeletal operation especially in the off-shore locations.

As a mode of protest, taking of oil workers hostage requires real commitment and tactics for it to succeed. So, it is a select group of agitators numbering about 200, most of whom are youths and had had a stint in the oil industry and possibly, in the military. In most cases, beautiful girls were recruited for the purpose of intelligence gathering and luring the oil workers into target areas. The squad also has an advantage of having hailed from the Niger Delta area, and having perfect understanding of its geography, particularly the creeks and the thick jungle. The majority of the members of this group of agitators are skillful swimmers too. They also guise as fishermen and boatyard workers. Contrary to the view of many a Nigerian who has condemned the act as lawless, accusing those involved as pirates or sea robbers since they demand for ransom, the author's investigators have revealed that, the original aim of those who initiated this form of protest, was to cause Shell to stop polluting their environment or forced the company out of the Niger Delta area by making the area hostile to operate. it was also gathered that off-shore location were chosen as the main target because their own intelligence reports indicated that, the state's ability to defeat them in such area was very limited. This is because most of the creeks are too narrow and shallow for the naval warships and boats to patrol. The thick jungle, which is one of the hideouts of the agitators, calls for guerrilla war professionals, should the state decide to attack them. That is, engaging in a guerrilla war with the inhabitants of an oil-bearing area whose major grievance is their call for Shell to obey the ethics of good oil field practice, and for the state to provide social amenities there. There are indications that, the state is ready to crush the group. However, the chances are that it will not only fight a losing war, but internationalize the struggle which will further dampen its image as dictatorial. In fact, some of the naval officers many of whom were from the Eastern Naval Command of the Nigerian Navy, who were deployed to quell the hostage crisis in the Nembe and Oproma oil bearing areas of Bayelsa in 1998 by then General Abubakar junta, who the author spoke to, admitted that the terrain was difficult to operate; and that it would require a full-scale force to overrun the hostage squad. No less than 70 workers of Shell, both local and expatriate staff have been held hostage but subsequently released by the squad as of May 1999. The release of the oilmen was not procured by the state military; nor was ransom paid. The deployment of naval men and warships to the Nembe and Oproma axis of Shell's location questions the rationale of the Nigerian state to go to war with its own people just because it wanted Shell to operate. Is the oil wealth not meant to better the well-being of the people who live in the host areas?

The demand for ransom before the release of oil workers held hostage was not part of the plan. The author's investigations showed that, some of the agitators who have hijacked the struggle for personal gains were not really part of the main hit squad. Rather, they are some of the frustrated youth who had no gainful employment, and decided to take advantage of the protests to make ends meet.

By and large, no one form of protest is capable of making any meaningful impact; rather, it is a combination of the various mode of demonstrations. Let us now turn to the responses of both Shell and the state to the reactions of the people of the oil areas.

6.6 Responses of Shell and the Nigerian State

6.6.1 Political Economy of the Responses

Though the Nigerian state uses its military might together with laws to support Shell's recklessness in the Niger Delta, they are not really in the best of friends. If anything, it is Shell that has tried to domesticate the Nigerian state, by taking advantage of the latter's weakness especially its technological dependence to pull through all hurdles placed on its path in the oil industry. As an alibi, Shell has always held the joint venture agreement responsible for the pollution of the environment in spite of the fact that it, indeed, produces the crude oil and violates the rules of good oilfield practice. Perhaps, it is only in Nigeria that Shell operates this way. The same buck-passing tactics have characterized the responses of both Shell and the state to the reactions of the inhabitants of the company's host oil areas.

In the thinking of the both Shell and the state therefore, it is the joint venture that should respond to the various forms of protests embarked upon by the inhabitants of the oil areas. It is rather difficult to expect the Agreement to respond moreso since it is not incorporated, and therefore, not a legal body. Rather, it is more or less, an agreed understanding between Shell and the state for the exploration and production of crude oil. In essence, it is the two major social forces directly involved in the production of crude oil and degradation of the environment, namely, Shell and the Nigerian state, that are responding, though they continue to blame the joint venture. The same is true of other foreign oil companies operating joint venture agreement with the Nigerian state. Some illustration is necessary at this point.

As noted, Shell's contention is that it really should not have anything to do with the agitation in its host oil areas moreso since it is only a technical partner/operator of the joint venture agreement. Suffice it to note that, the company should not be held responsible for polluting the environment. Nor should it be expected to provide social amenities in its host oil communities especially as it pays taxes to the state. Again, it is the joint venture that should do all that. Shell sticks to all this when the agitations is not fierce. By the time the violence has halted its operations, especially when acts of sabotage and prolonged communal protests as evidenced in Ogoniland and Isoko oil areas have caused Shell to loose million of barrels of crude oil and revenue, it jettisons its earlier position, and begins to count its own direct contribution as if it no longer produces crude oil for the joint venture. At the end, it heaps the blame on the state, forgetting that its operations caused the crisis. Shell has got the variety of double talk across the oil areas in the Niger Delta. For instance, it was not until the MOSOP gave Shell's relative reckless operation in Ogoniland a world-wide publicity, that the company reluctantly owned up to its portion of the wreckage, blaming a larger part of it on the Nigerian state. Buck-passing usually provided an opportunity for the company to publicize the catalogue of all contracts it had awarded to notable indigenes of the area aimed at pacifying the people and providing them social amenities; and scholarships awarded to the indigenes, among other cosmetic issues. Furthermore, Shell's clandestine action and double standard was unearthed when the MOSOP, individuals and other groups sympathetic to the Ogoni cause, provided overwhelming evidences of its complicity in financing, procuring arms and providing logistics for then Colonel Okutimo-led invasion of Ogoniland. Shell's denial of covert connivance with the state to crush the uprising in Ogoniland was, as ex-staff of the company put, to protect its image internationally, passing the buck back to the state will hardly help Shell to redeem its battered image in Nigeria.

The only difference between other oil companies and Shell is that, some of them like Agip with a relatively smaller concessions, decided to be friendly with their host areas by providing social amenities there, though still degrades the environment. In fact, Agip cannot afford to behave like Shell in terms of sponsoring conflicts, or else it is forced out of its locations by the agitating youths. Shell's reaction to the response of Agip to its host oil areas has been that, it would raise lost of product.

On its part, the state has, through the NNPC, first decided to strike an agreement with Shell and other operating oil companies like Chevron to jointly credit the Corporation and the company for whatever projects like building market stalls that they undertake in the host oil areas rather than make it look as if it (state) was insensitive to the plight of its own people. It is all part of the plot by the state to convince the people who live in the oil areas that it is in partnership with the oil companies in providing them social amenities. Hence, signboards conspicuously displayed on community projects carry captions like: "Market stalls are provided by NNPC/Shell joint venture". Pertaining to the degradation of the environment, the position of the state is that, as the technical operator of the joint venture, it is Shell that should be held responsible. Its contention is that, if Shell were environmentally cautious in its operation, there would be less cases of pollution as currently experienced in the host oil-producing areas in the Niger Delta. But it is more of a radical defensive position since it supports the company's recklessness. Having clarified the political context of the responses of both Shell and the state, let us now begin the real discussion, starting with Shell.

6.6.2. Shell

In 1971, Nigeria joined the Organization of Petroleum Exporting Countries (OPEC). And in line with the OPEC's resolutions of 1968, the Nigerian state intervened into the upstream sector of its petroleum industry through the joint venture. Before then, Shell and other operating oil companies had dealt directly with the inhabitants of their host oil areas on matters of community development, land and compensation through their chiefs and other influential indigenes who claimed they represented their people. In fact, they are no less than part of the conflict merchants in the areas earlier referred to . Top on the agenda of Shell then, was the maintenance of relating peace that would permit oil production to take place there. Through the conflict entrepreneurs, some measure of peace was assured in the area. It was not an altruistic duty, for Shell paid them in the form of gratis, employment of their wards and family members, and award of scholarships to their children. This was part of Shell's plot to build up a clientele of traditional rulers in the host oil areas who are to brief the company on security matters there. That is not all, for until the Land Use Decree of 1978 came into force, Shell was able to settle compensation for land acquired for oil exploration, production, transportation and storage. This was in spite of the Petroleum Decree of 1967 which still contained the clause of sole ownership of land on the state as enshrined in the Mineral Act of 1914. Which means that both Shell and the state were silent on this aspect of the Decree since it made relationship with the host area relatively friendly. Though Shell was the only foreign oil company that had and still has its own police force called shell Police' it still relied import on the chiefs/traditional rulers for pacifying the people in the oil areas in the 1960s and early 1970s agreement is that, since they are nearest to the people they are in a better position to calm them. Not only that. As officials of Shell had put it, 'the initial idea of the company having its own police was basically for light security purposes in its quarters, officers, quality control centres, but not for its locations, platforms and terminals.[19] There was really no threat to Shell's operation at the time due in part to the lesser level of pollution and ignorance of the people on the implication of gas flaring for their health and increment. In fact, the flare was seen as an alternative to electricity by most of the people. All this explained why most of the chiefs/traditional rulers had privileged access to the company which they exploited for their personal gains. Suffice it to note that, long before now, most of the linemen of Shell had been colluding with the company to underdevelop the people who live in the host oil areas but unknown to them.

However, it was part of the findings of the Presidential Task Force on the Oil Areas of 1988, with Nura Imam, an air vice marshal and then Minister for Mines, Power and Steel, and Chairman, that unearthed a tip of the deep-seated corruption in the oil areas with their chiefs and traditional rulers implicated in the scandal. They were found to have received from Shell and other oil companies gratis and monies to settle compensation claims, which were never paid, and contracts awarded to build roads, market stalls for which money was paid, but no work was done. It was gathered from the Bendel State Chapter of the Task Force that Shell, Elf, Agip and Gulf (Chevron) had, between 1980 and 1986, paid about N500 million to their fronts like chiefs and traditional rulers for unexecuted contracts awarded for the provision of social amenities like road, market stalls in their various communities. The culprits in Kokori/Orogun Community for instance, owned up that they too had collected about N30 million from Shell for the period, 1980 - 1986, through they could not account for the money.[20] The only sanction against them by the community was that, they would never negotiate with Shell on behalf of the people. As to be expected, the state did not prosecute them. Nor did Shell deny paying them. That was and still the trend throughout other oil-producing areas making it unnecessary to embark on a blow-by-blow account of other cases.

It was due partly to the findings of the Presidential Task Force on the Oil Areas that further angered the youths to sack the leader of the various movements and take over struggle. They felt betrayed by their leaders. As a consequence, anti-oil companies' demonstrations began to deepen in the late 1980s. In response, oil companies have decided to reduce using some of the influential indigenes of the host areas as fronts. Rather, they prefer to provide some of the basic social amenities directly to the communities amid their continued pollution of their environment through gas flaring, spillage and dumping of untreated drilling wastes. A move that was resisted by the petroleum whose one major source of income was undermined. But the oil companies did not yield to them That brought more unrest into the area as the linemen began to move against the oil companies. In the case of Shell, its trend of response over the years, is that, unless an oil community agitates and does so very violently, nothing is get done. Past actions of Shell have shown that agitations must be capable of disrupting the exploration, production and transportation of crude in order for Shell to really listen to the demands of the people of the affected community. It does not mean that Shell would act immediately in terms of stopping pollution and providing social amenities. It was for this reason that Shell completed a full cycle of oil operation in the Oloibiri field in 1980, leaving the Opume, Otuogiridi and Otuebuyi oil communities worse-off than what they were environmentally, and socio-economically partly because the company was not seriously confronted by the people. Also, that explained Shell's delayed response in Ogoniland, which compelled the MOSOP to force Shell to halt its operations there so as to avoid a repeat of the Oloibiri oil areas, experience. That is not all, for unless protests are well organized and protracted, the tendency is for Shell to wait for tension to calm and thereafter, resume operations. Again, it is the MOSOP's prolonged resistance against Shell that forced the company to pull out in 1993. Some specific reference is important at this point.

As explained earlier, the Isoko oil areas are about the most restive in the West of the Niger delta. They have been worst hit by ecological crises and depending rural poverty as a result of the activities of Shell with the exception of Irri where Agip operates. Shell is the dominant oil company operating there since 1958. Past records have shown that hardly any year had passed without Shell spilling crude oil. For 40 years, too, Shell has been flaring gas in both day and night there. The spillage that occurred in the Uzere community in River, other streams, lakes and traditional fish ponds of the area. As usual with Shell, there was no clean-up; nor any environmental and social economic impact assessment conducted in the area. Nor did the state. Deprived of light, water, road etc, and coupled with years of incessant pollution, protests against Shell in the area cannot but be violent and protracted. By 1991, it was obvious that pent-up grievances and regular outburst of the people showed that Shell's frontment could no longer guarantee peace that would permit further exploitation of crude oil there. For instance, it was revealed by the IYM that, though Shell part-financed the new palace of the Ovie of Uzere, Odhime Udogri I, gave him N30 million as gratis, and linked him with Seismograph Services Limited for which he acted as a foreman, was awarded N30 million contract to tar the 20 kilometers road linking the Uzere community with other area in Isoko area, got paid and never did the job, he could no longer pacify the youths.

Again, OMPADEC's local contractor, Opia Ventures Limited, which bought one 2.5 KV generator for the Ovie as gratis, did not help matters. On its own, OMPADEC's portfolio contracts valued at N30 million awarded to him and for which he was paid and no work done, failed to calm the nerves of the agitated youths.[21]

Angered by the collaboration between Shell OMPADEC, and the Ovie and his cohorts, the Uzere Youth Movement, a chapter of the Isoko Youth Movement, sacked him in June 1994, and was subsequently forced into exile. Shell's operations in the area were halted and still suspended by 1996. Though tension was reduced these, there was no signs that Shell was really back to begin production. In fact, it is Isoko Youth Movement that is currently taking Shell's workers hostage in the troubled area of Delta State. As of July 1999, 67 workers of Shell were held hostage, with seven of them expatriate. They were later realised after weeks of incarceration. This and other protests have compelled Shell to threaten to relocate its Warri office to Port Harcourt. But it is not safer there for the company.

Still on Shells half-hearted response. In 1979, 500,000 barrels of crude oil spilled from its storage tanks at the Forcados terminal. A formal protest was lodged with Shell by representatives of the Forcados area most of whom resided in Warri. This was so because those directly affected by the pollution were too, poor, ignorant and had no access. After prolonged negotiations, Shell paid N1.5 million as damage to the people of the affected community. It was not really clear what indices were used by Shell and the State in computing the claims. Which means that the amount was a far cry from the damage suffered by the people estimated at N150 million by independent surveys conducted by a faction of the Ijaw Youth Movement. Worst still, the money was reportedly confiscated by some of the representatives of the people, some of the officials of the Lands and Estate Department of Shell, and those of the State's Ministry of Lands and Housing who were all involved in the calculation of compensation. It was found out that no consideration was given to the assessment of the damage done to the environment and restoring normalcy to the social and economic activities of the people.[22]

Shell is not alone in providing cosmetic social amenities, other oil companies equally do the same. Mobil for instance, built a N30 million Health Care Centre at Eket. Road and portable water valued at N30 million were provided by Agip in Delta and Rivers States. However, unlike Shell, Mobil, Agip and Elf among others, pre-emptied protests in their host communities by providing some social amenities for the people who live there amid the degradation of their environment. All has revealed the hollowness of the oil companies' so-called development programmes in the Niger Delta area.

Amid the failed cladenstine dealings of Shell with chiefs/ traditional rulers and other local allies as unrest deepens in its host oil areas, there are indications that the company might be force to rethink its response in order to guarantee the safety of its workers and oil installations. Alone, it will be difficult for Shell to succeed. Perhaps, with the state, it might be possible. As noted, because Shell suspects the state's intention in terms of pollution and compensation matters, there is very little chance for success for a joint action. In the interim, Shell still sticks to its guns that it has no business improving the material well-being of the people who live in its host areas, a position the state is bent to force the company drop. More details on Shell's response later. Let us now discuss the response of the state.

6.6.3 The State

With the introduction of the joint venture into the upstream sector of the petroleum industry, the state has, in theory, roped itself not only into the payment of oil exploration and production expenditure incurred by Shell, among other oil companies, but also, the payment of damage, compensation and community projects hitherto borne by the company. Since the state managers are not really interested in the development of the oil areas and the enhancement of the material well-being of its people, it is not surprising that its response has sought accommodation with the ultimate interest of Shell in the Delta. Land, on which the subsistent economy of the area hinges, is exclusively owned by the state. Which means all land where crude oil is explored, produced, transported and stored, does not attract rent since the Land Use Decree of 1978 came into force. Because more oil wells have been found since 1978, it means more land has been taken up by the state and for which no compensation was paid. With the battle over the ownership of land won by the state at least in Thenya Shell had more freedom to operate and even recklessly, but all is to the detriment of the people. This explains why the state does not consult those who live in the oil areas before fixing the rate of compensation paid for crops that are destroyed in the process of oil exploration and production. For one thing, they have been turned into squatters in their own ancestral land. Worst still, how much a farmer gets paid as compensation for his crops, was dependent not so much on the list of rates as shown in Table 6.5. Rather, a lot has to do with his connections with both the Estate and Land Departments of

Table 6.5 Selected Crops and Economic Resources and Rates for Compensation in the Oil-Producing Areas

Source: Author's field works: Ministry of Lands and Survey, Delta, Edo, Rivers States; Shell's Estate Department, Western Division, Warri; Eastern Division, Port Harcourt 1990; Community files of oil communities in Delta and Rivers States, 1997.

Note: Some of the lawyers handling compensation suits for the people for the oil area confirmed to the author that this rate was not sacrosanct. One of them, one Prosper Ezo, said he usually secured higher rates from Shell, which means there is no standard rate.

Crops and Economic Resources	Categories and Rates		
	Mature	Immature (2 Years to bearing age)	Small/ Seedling (1-2 years)
Beans (Mature)	13k	6k	-
Bitter Leaf	9k	-	-
Cassava (plant)	15k	6k	-
Groundnut	9k	11k	-
Maize	25k	11k	-
Okro	39k	13k	-
Pepper (Hot)	22k	13k	8k
Plantain	39k	16k	8k
Sugar cane	33k	17k	-
Tomato	44k	13k	-
Yam	17k	9k	-
Bambo (indigo)	N2.20	-	-
Oil Palm/tree (maintenance)	N6.50	N3.30	N1.50
Native peer/tree	N5.50	N1.65	88k
Raphia palm/tree	N8.50	N3.85	N1.65
Rubber/tree	N1.37	83k	22k
Agbono tree/one	N3.30	N1.10	-

Table 6.5 (cont.) Selected Crops and Economic Resources and Rates for Compensation in the Oil-Producing Areas

Crops and Economic Resources	Categories and Rates		
	Mature	Immature (2 Years to bearing age)	Small/ Seedling (1-2 years)
Timber trees (iroko, abura ironwood etc):		N8.00	-
- Over 1 meter	N8.00k		
- 60 cm 1 meter	N4.00	-	
- Below 60 cm	30k	-	-
Water sights:			
- Fish pond	N500.00	-	-
- Lakes (private)	N600.00	-	
Fishes (determined by the numbers, size, and weight) 100kg and above	N300.00	-	-

both Shell and the state. Past evidences have shown that lawyers, who defend the aggrieved people, have also taken advantage of the ignorance and poverty of these people to appropriate compensation monies. Some illustration is in order here.

As explained earlier, it was the finding of then Presidential Task Force on the Oil-Producing Areas, former of OMPADEC, that the majority of their inhabitants became aware that Shell and other oil companies like Agip, Elf had paid compensation for crops destroyed either during oil exploration and production or spillage. Before then, it was classified. In essence, those who act as a go-between for the oil areas, the state and oil companies are doing so because of personal gains. For instance, in the Isoko, Ughelli, Ndokwa and Ethiope oil-producing areas in Delta State, the author's investigation showed that, Shell paid about N15 million for the period, 1983 and 1988. For the Isoko areas in particular, Shell paid about N3 million; yet, it was unknown to the majority of the people except the Ovie of Uzere and the company's other pointmen. When then Bendel State chapter of the Task Force planned to visit the area, they became jittery. Afraid of a possible leakage to the Task Force by the aggrieved people, an *ad hoc* meeting was held by the chiefs/traditional rulers who received the money. based on personal contacts and family ties, and pretending to be as if Shell just paid compensation for damaged crops, 100 farmers were hurriedly selected from an estimated total population of 200,000 farmers for payment with each receiving N5,000 as compensation, promising to pay them more as soon as other cheques were cleared. Though the compensation was received with mixed feelings by the rented farmers, they could not bare their mind to the Task Force team when it visited. This was because the so-called linemen shielded them away from the Task Force. A different crowd was hired to meet the Task Force, who expressed gratitude to the state for its help. That is not all, for N5,000 was a far cry from the value of crops destroyed which was estimated at N600 million over the period under review.

In Ughelli oil areas, about 30 rented farmers were paid N3,000 each. Forty farmers got N2,000 each in Ethiope area. Worst still, over four lists of rates and beneficiaries were kept by the frontmen and officials of Shell and the state, all aimed at enriching themselves at the expense of those who really bear the brunts of oil exploration and production. For instance, those collaborators in Shell's Estate and Land Department have a ready list of beneficiaries numbering over 50,000 in the Isoko, Ugelli, Ethiope, and other oil areas in the Niger Delta. So, too, are their cliques in the Land Ministry of the host state governments. It is from this list that compensation monies are disbursed to the majority of the ghost beneficiaries, monies which they later collected. Both Shell's management and the officials of the state are aware of all this, but decided to allow it continue because it profits them. Shell is most guilty of this, but obviously not the only company involved. In fact, it so pervasive across the entire oil-producing areas in the Niger Delta that a community by community case study is not necessary.[24]

However, it is important to note that drive for selfish gains by the frontmen of Shell among other oil companies, and the Nigerian state have depended intra-communal clashes and anti-oil companies protests in the Niger Delta. Realizing that even their own people now collude with officials of the oil companies and the state to exploit them, the youth wing of most of the anti-oil companies movements have lost confidence in their leaders as explained earlier. Where the youths have sacked the leadership of the movements, it is marginalized. All this explains why the youths have remained ever restive and committed to sustaining the struggle alone. The youths are not only more daring in their protests, but also, better coordinated and have a network of intelligence gathering and groups of consultants on oil exploration, production and transportation. In response, the state has became more aggressive.

Thus, across oil/gas locations in the oil areas of the Niger Delta, the army and navy are on guard, with the airforce kept on the alert. But their presence in the host oil areas appear not scared the agitated youths who seemed to have vowed to fight on until Shell, other oil companies and the state have stopped degrading their environment. Shortly before the Ogoni Bill of Right was launched in 1991, the youths of Umuechem in Rivers State had, in November 1990, called for the exit of Shell from the oil-rich community following the destruction of their environment and economy since 1957 when the company began operation there.

Their complaints:

What is a secondary school scholarship of N8000 given to 10 of their indigenes compared to billions of Naira that the state and Shell have realized from about 36 oil wells in our area? Every family in Umuechem has a manual garri processing machine. O what use therefore, is Shell's manual garri-processing machine? The water from the bore-hole provided by Shell is untreated and the tap does not even flow regularly. So, what does it mean that we are insatiable? Even we prefer to fetch water from the Otamiri River, but it is polluted by Shell. What can be worse than that?[25]

It was not mere complaint. For, the youth actually barricaded all access roads leading to Shell's locations in the area. Shell, in turn, reported the case to the state. The then military governor of Rivers State, Godwin Abbey, an army colonel at the time, ordered the deployment of troops to the area. 500 policemen first moved into the town, with plan to send in the 300 soldiers later. By the early hours of November 1, 1990, the village was razed down, leaving A.A. Ordu, the King of Umuechem and 50 others dead. The town was occupied by the policemen for one week during which they looted their homes, and killed goats, fowls and sheep for meals. Three policemen who were declared missing in operation were later found beheaded by the irate youths.[26]

In no other area has the state declared a full scale war against the youths than in Ogoniland. In fact, it was not surprising that the state sent its guerrilla-trained officers and men to dislodge and kill youths of the MOSOP who were bent on chasing Shell out of the area. Though no accurate figure of casualty was recorded, no less than 600 youths and 400 children and adults were killed by the military in Ogoniland.[27] This was in addition to the hanging of Ken Saro-Wiwa and eight other Ogoni environmental activists in November 1995. As of the time of writing, the army was yet to fully pull out of the area. Nor has Shell began operation there. A repeat of the scale of shooting and killing reminiscent of the Ogoni during which over 1000 people laid dead, was carried out after the Kayama declaration in December 1998 by the Ijaw Youth Movement. Like the Isoko Youth Movement, those of the Ijaws wanted Shell out of their area as the company still sticks to the pollution of their environment. Though they were subdued by a combined team of the army and navy, Shell's installations are still being vandalized and its workers held hostage in Bayelsa State as of July 1999.

The state's deployment of its armed forces to kill its own people in order to make the oil areas safe for Shell, among other oil companies, to operate seems not to be yielding result as agitations deepened and the youths better armed than before. As a result, the state created OMPADEC to replace the Presidential Task Force on the Oil Areas. As a political outfit, the Commission is meant to play carrot and stick politics for the state in the Niger Delta. So, it is not really concerned with the desired development of the oil area contrary to the view of many a Nigerian. As part of its politics, it believes in awarding portfolio contracts as a means of building up the needed political support in the area for the Babangida junta in need of civilizing itself. That explained in part why only N65.4 million out of the total of N4.8 billion of contracts awarded in

1993 and for which payment was made, was accounted for. Across the oil areas of the Niger Delta, therefore, there was (and still) complaints and grievances that then chairman, A.K. Horsefall, a state security expert, awarded contracts to those seen to be allies of the state in the area, and others who are government officials' fronts who are from the oil areas. He did what the state wanted him to do. The Bonny Oil Community development Association think otherwise, accusing him of woeful failure. In an open letter to him, they complained:

> ... One thing appears clear now: OMPADEC under your leadership has so far, either by omission or commission, decisively marginalized the development interests of the actual oil producing communities in Bonny Local Government Area of Rivers State. Your public notice advertised on page 13 of *Daily Time* of Friday, December 31, 1993 not only dusts facts when compared with on-ground realities in Bonny LGA (and other oil areas), it graphically tried to cover up the brazen dehumanization, unfairness and inequality that characterized your Commission's management of the 3% Oil Communities Amelioration Fund ...[28]

The position of the Bonny Oil Community development Association was similar to that taken by most of the actual oil-producing communities across the Niger Delta. But I think it is mistaken position afterall. For, some of the leaders of these associations may not be really different from the then Chairman of OMPADEC since the Commission became yet another avenue created by the state for selfish political gains and material acquisition by its managers and their cronies. The N3 billion debt which the Commission owed its contractors most of whom are either government officials or their fronts, is in part a demonstration of its hollowness. All has explained why the replacement of A.K. Horsefall by E. Opia, who spent about two years as the Commission's boss before he, too, was replaced by P. Omatsola, a retired navy officer, has done little or nothing to make it responsive to the development needs of those who bear the brunts of oil exploration and production. Even the Niger Delta Development Commission Bill currently submitted to the National Assembly by the Obasanjo government, which, if approved, will replace the OMPADEC, may not be any different. This is because the state still has not undergone a rethink of its mode of surplus extraction in the country's petroleum industry and its conception of development particularly the rural areas.

6.5. Concluding remarks

Generally, this Chapter has demonstrated that the deepening degradation of the environment for the host oil-producing areas to Shell, has worsened the incapacity of those who live there to reproduce themselves socio-economically and culturally. And that Shell, though most guilty of the violation of good oilfield practice, other oil companies are not saints. Rather, their scale of recklessness is relative less than Shell's moreso as they have smaller concessions. The contention is not that Shell and other oil companies cannot be made to obey the rule and regulations governing oil exploration and production if the Nigerian state so desires. In other words, Shell is able to degrade the environment of its host oil areas in the Niger Delta with the backing of the state. Insightful as this Chapter has been, space did not permit us to elaborate more on the activities of Shell and the state on specific communities. It is for this reason that the next two chapters are devoted to the case studies of the Ughelli and Isoko oil communities aimed at bringing into greater, the depth of population and rural party in the area. The Oloibiri oil area, where Shell had completed the full cycle of oil exploration and production, was a subject of another study conducted by the author. References will be made to it to reinforce my argument.[29] Let us turn to the discussion on the Ughelli oil areas.

Notes and References

1. Walter Rodney, *How Europe Underdeveloped Africa* (Dar-es-Salaam: Tanzania Publishing House), 1972.
2. Frantz Fanon, *The Wretched of the Earth* (New York: Grove Press), 1969.
3. See J. Iloeje, *Geography of Nigeria* (Lagos: Longmans), 1974.
4. based on the author's field trips to these oil-producing council areas between 1987 and 1994.
5. *ibid.*
6. Based on the author's numerous field trips to these oil-producing communities between 1987 and 1995.
7. *ibid.*
8. *ibid.*
9. based on author's field trips; and data obtained from communities' files.

10 For details on the study, see Sogo Awobajo, 'An Analysis of Oil Spill Incidents in Nigeria: 1976-1980', in *the Petroleum Industry and the Nigerian Environment*, (Lagos: NNPC) 1982.

11. *ibid.*

12. Abstracted from the various documents obtained from the community files, Shell and NNPC.

13. Based on the author's trips to the area, 1989-1993.

14. *ibid.*

15. *ibid.*

16. See Suit No. HCO/18/90, Ivbrogbo Community vs. Shell, High Court Oleh, September 1989.

17. See Suit No. SCK/21/81, Chief S.O. Ubini vs Shell, July 19, 1985; and others like Suit No. IV/89/71, Warri High Court, 25, November 1971; and Irri Community vs. Agip.

18. For figures on Shell, see, *Employee Information*, Lagos, Shell, April 20, 1994, and data for other companies were obtained from their office and field staff at Lagos, Warri and Port Harcourt, 1994 and 1995.

19. Based on the author's interview with officials of Shell in Lagos Office, 1993.

20. Author's interviews with L. Omoru, then Commissioner of Commerce and Industry, Bendel State, and Chairman, Presidential task Force on Oil Areas, Bendel State Chapter, in 1989; and the youths of the area in 1990.

21. Based on author's interview with tie IYM, A. Akoko, Chairman of the Reconciliation Committee on the Dispute; and some staff of Shell's office at Warri, 1991, 1992, 1994 and 1996.

22. Based on author's checks with officials of the Ijaws Youth Movement at Warri in 1988.

23. based on the author's checks with the Isoko Youth Movement in 1992.

24. Based on the author's field trips to the areas.

25. based on the author's field trip to the area in December 1990 shortly after the conflict. See, also, *African Concord*, 1990, pages 23-27.

26. *ibid.*

27. based on the author's checks from field trips.

28. See, *The Guardian*, January 26, 1994, pp.16-17.

29. See, Daniel Omoweh, *Political Economy of the Operations of Shell Petroleum Development Company in the Oloibiri Area of Nigeria*, (Stuttgart, Germany: Bread for the World), 1998.

Chapter 7

Shell, Environmental Pollution, Culture and Health of the Ughelli Oil Communities

7.1 Introduction

This chapter establishes the linkage between how Shell explores oil, and produces crude oil, the culture and health of the people in one of its host oil areas, the Ughelli oil communities of Delta State. Within this context, it looks into the process of the degradation of the environment of the area through the company's seismic operations, indiscriminate dumping of untreated dangerous drilling wastes on land and into swamps, flaring of gas, seismic operations and frequent oil spillage; and how these forms of pollution have deepened the incapacities of those who live there to practice traditional medicine because of the destruction of, and resultant scarcity of herbs with therapeutic values. It also examines how Shell's incessant violation of the cultural beliefs of the people through the destruction of their traditional places of worship and scared sites, killing of their totemic animals like python by the company's workers have inflicted serious damage to the social fabric and culture of the communities thereby restricting their quality of life and upturning the traditional medical system. Finally, the chapter addresses the reactions of the people in the area to capitalist exploitation, and how both, Shell and the state, responded to protests. It evaluates the overall effects of the capitalist development promoted by Shell and the state in reproducing the underdevelopment of the host oil area in particular, and the country generally.

As noted, the nature of the underdevelopment of the host oil producing communities across the Niger Delta inclusive of the Ughelli oil areas, has been largely understood in the context of lack of social amenities such as pipe-borne water, tarred roads, and hospitals which both the state and Shell are expected to provide in reciprocation for crude oil extracted from there. Experiences of the host oil areas have shown that even when Shell provides some of these amenities, they only facilitate the exploitation of the communities as evidenced in the construction of its access roads that link up its various oil and gas fields, but not necessarily meant to develop the host communities. On its part, the state is not doing anything serious to redress the imbalance. This chapter does not dwell on this aspect of underdevelopment of the oil-bearing areas because other authors have covered it considerably well[1]. Therefore, it is a departure from existing works in the field by focusing on the crucial but neglected dimension of the pillage of the area. That is, the consequences of environmental (oil) pollution on the culture and traditional health system of the people. As part of the overall hypotheses of this study that, society which is always in constant motion cannot be understood outside the specific state of development of its production forces and the corresponding social relations of production, it contends that the culture, especially the traditional beliefs system of the people in the Ughelli oil communities, so embodies their traditional health system that its distortion and destruction invariably hinder the health of the inhabitants of the area.

The nature of the underdevelopment of the area is also understood from when it first had historical contacts with the forces of British colonial capitalism. This is because they had initially distorted the traditional religion and native health care system of the area, which paved the way for the subsequent production of crude oil by Shell and the state finally destroyed its cultural development continuum.

The Ughelli Local Government Areas consist of two Councils which were created out of the former Ughelli Local Government Area when Delta State was split into 19 local government areas on 27 September, 1991. While the Ughelli North Council has its headquarters at Ughelli, South Council is administered from Otu-Jeremi. The population of the Councils 161,157 (North) and 131,291 (South)[2].

The culture and traditional health care of the Ughelli oil communities are not strictly different from those of other seven local government areas in Urhoboland. In discussing the impact of environmental pollution on their culture therefore, reference will be made to these other oil-bearing areas only in so far as it enriches or clarifies an argument. The study is also limited to the upstream sector of the petroleum industry. That is, concerned with the activities of Shell and the state in the exploration, production, and transportation of crude oil and natural gas in the area. Although the oil servicing companies like Seismograph Services Company largely explores crude oil for Shell in the area, it does not constitute a separate focus of study. Rather, together with OSCs, they are discussed as part of Shell's capitalist mode of operation.

The Utorogun gas plant is essentially more in the downstream sector then the upstream of the country's petroleum industry. As a result, it will not really be discussed as part of the activities of Shell in spite of the fact that the gas plant was jointly built by Shell and the state, and even located in the company's oil/gas field in the even. Reference will be made to it to clarify an argument. In a situation where the company's oil and gas field has about 23 oil wells as in the case of Kokori/Orogun, particularly when the former has five wells and (Orogun) with 18 wells but both of them are within Urhoboland, no real distinction will be made; rather, they are discussed together.

Most part of the Ughelli North and South Council areas are drained by the Forcados River and its tributaries especially the Okpare creek, Oguname creek, Ewu creek and that of Aramaire. It has two distinctive seasons - dry and rainy. The average rainfall is about 266.7 cm and temperature is 80°F. The vegetation is mangrove swamps and forest. The major occupations of the people are farming and fishing. Traditional medical practice is also an important vocation because of the wealth of the area in herbs. The area is rich in both crude oil and natural gas. Shell accounts for over 95 percent of oil exploration and production in the area. Ughelli town was the second location where crude oil was discovered in large and commercial quantity in 1958 by Shell in the Niger Delta after the Opume, Otuatagi and Otuogidi oil areas of the Oloibiri field in 1956. The largest gas reserve in the country also located is in the area; that is Eruomukowharie gas field which compelled Shell to build the Utorogu Gas Plant. Notable villages and towns in the area where oil and gas fields are located are: Orogun, Afiesere, Eriemu, Evwreni, Eruomukowhairie, Otu-Jeremi and Umolo. By December 1992, Shell's oil wells in the area number 70 consisting of Afiesere (8), Eriemu (14), Enwreni (5), Koroki/Orogun (22), Ughelli East and West (11), Utorogun (10)[3],

Culture has many connotations; but in this study, it means that complex whole which includes knowledge, belief, art, morals, law, custom and any other capabilities and habits acquired by members of society; which is restricted here to the traditional religion of, and medical practices native to the people of the area[4].

Traditional religion, under the broad definition of culture, is used here to mean a man's effort to satisfy certain needs, including emotional and psychological ones, by establishing and maintaining cordial relations with the supernatural through their ancestors among other divinities[5].

Traditional Medicine, on the other hand, means

> "the sum total of all knowledge and practices whether explicable or not, used in diagnosis, prevention and elimination of physical, mental or social imbalance and relying exclusively on practical experience and observations handed down from generation whether verbally or in writing" (WHO, Afro. Rep. 1976:1-3)

This definition is quite all-embracing. But given the nature and objective of British colonialism, traditional medical skills were discouraged and not even allowed to be recorded in written form. The most viable form of storing knowledge on traditional medical skills therefore, is still through incantations, ritual, songs and cult procedures which are memorized by the practitioners. This is how traditional medicine is practiced in Urhoboland storage of knowledge of native medical especially through the stills[7].

As noted earlier, British rule preceded the advent of Shell in the Niger Delta. Again, though British rule had disrupted the development continuum of the colony, its input varied in degree. It is for this and other reasons that the role of British rule in the destruction of the cultural environment of the Ughelli oil area is discussed.

7.2 Foundation for the Destruction of the Cultural Environment of the Ughelli Oil Areas: An Overview of the Role of British Rule, 1900-1960

In this section I present briefly, the role of British colonialism in setting the stage for capitalist by the early corporate trading companies, on which the foreign oil-producing not the least, Shell, began its operation there. To be able to have a fuller understanding of the extent of destruction that the activities of Shell have done to the cultural environment of the area, a brief discussion on its pre-capitalist period is first undertaken.

7.2.1 Pre-Capitalist Cultural Environment

The pre-capitalist cultural environment is restricted here to mean the traditional religion and traditional medical system of the people in the host oil-area. Let us begin the discussion with traditional religion.

The people of the Ughelli oil communities are very religious. They believe in and place high confidence on the supernatural being, often expressed as *Oghene*, meaning God. Their conception of the Supreme Being is that, it is the overall owner and controller of the whole universe, that is, *Oro vw'akpo* (Supreme Deity). He (Supreme God) is predominantly worshipped both directly and indirectly. The direct mode of worship is either done through personal requests to God, or through the male head of the family. For instance, the head of the family, whether confronted with danger or not, worships the Supreme Being every morning with a quantity of *Orhen*, meaning native white chalk, which he keeps in his left hand and breaks into powder form with his thumb. While doing this, he addresses *Oghene* in his generic names such as *Oghene Osonobruwen*, and makes his heart desire to Him while looking up into the sky.

The indirect mode of worship on the other hand, is done through *Edjo* and *Erhan* (divinities) accepted by the people as their ancestors and ancestresses whose power is derived from the Supreme Being. The people are united under the divinities. The *Oghene-Egodo*, god of the compound, is symbolized by the *Oghriki* (New-bouldia) tree, which serves as the medium through which the Supreme Being is communicated with. The choice of *Oghriki* tree, a tropical plant, is because it attracts various spiritual forces. He uses palm wine, white kolanut and white cockerel to serve the god. The white colour symbolizes that God is pure and also manifests the people's mind at worship.

The worship of divinities is confined to the socio-political group that accepts their reality. This, does not however, mean their influence is not extended to devotees who live in the areas outside of their immediate scope. For instance, the power of *Ekene* is restricted to the sons and daughters of Agbarha who are both at home and abroad. The same is true for *Edjo-Orhiara* for the people of Orogun, and the *Ogba-Urhie* for the people of Jeremi and some part of the Udu clan. Other are the *Umanoku* for the people of Ugheli; and *Igboze* for the Olomu' clan. These divinities were believed to have served and still do as guardians on questions of war, fertility, and ethical purposes. A lot of taboos are associated with these divinities, explaining why festivals are celebrated in their honour. It is also in these festive periods that contacts are renewed with the divine forces. Herein lies their sociological importance to the culture of the people. The Usiefurun people for instance, celebrate the *Ogba-Urhie* festival not only to mark their divine origin, but also, to honour the python, the sea goddess believed to have guarded and guided them in their migrations from the Bosrin Kingdom through the sea to their present location in Delta State. The python is therefore, a totemic animal and honoured with festival. Table 7.1, for instance, shows a summary of the major communities in the area and their revered animals.

Table 7.1. Communities and Some of the totemic Animals Under the Ughelli Areas

	Community	Totemic Animal
1.	Ughelli, Afiesere and Ufuoma	*Orhua* (Antelope)
2.	Ughievwen (Jeremi)	*Isekpe* (Snail)
3.	Orogun	*Ogborigbo* (Iguana); *Erhako* (Dog)
4.	Evwreni	*Ararevwa* (Crocodile); *Ewherin* (monkey)
5.	Olomu clan (Okpare, Umolo)	*Edjerhe* (Crocodile); *Ohorhe* (Fish);
6.	Usiefurun (Jeremi)	*Ogbo* (Python)

Source: Author's field work, June 1993.

As it will soon be made clear, these and other communities are incidentally the major oil-producing areas being underdeveloped by both Shell and the state.

The *Erivwin*, cult of ancestors, is strongly believed by the people to know all that is happening either in hiding or in the open. The concept of the dead by the inhabitants of these oil areas is that, spiritually, they are still keeping watch on them who are living to avert evils and punish evildoers in order to bring harmony in society. They (ancestors and ancestresses) are worshipped at anytime of the year but elaborate ones are usually done in periods of annual traditional festivals.

Erhi, meaning man's self-predestination among the Urhobos, guides a man in the direction that enables him to actualize what he is destined to become in life. *Oghene*, the Supreme Being, is believed to have played a crucial role in the *Erhi* of a man. It is therefore, an important component of the religion of the Urhobo people.

The Urhobos have a traditional four-day week with each of them meant to serve for specific spiritual purpose. *Edewo* (main market day) and *Eduhre* (small market day) are believed to be sacred days for serving the divinities and ancestors. It is traditionally believed by the people that, on those days, the spirit of the ancestors, converge and is therefore, an abomination for the living to see them physically. Rituals to appease these divinities are also done on such days. *Ediruo* and *Edebi* are working days. The four-day week and the two major seasons, dry and rainy, are of traditional significance for fixing planting, harvesting and festival dates. For instance, the rituals for the divinity of prosperity are performed in the dry season when the rivers, creeks and streams are believed to have receded with the various divine marine spirits.

The traditional belief system of the people is as important in understanding and explaining their native religion as it is in having an in-depth analysis of their traditional medical practice. The traditional medical system of the area is basically rooted in the people's native beliefs. Its major objective is to prevent and treat diseases in order to save life. It also brings harmony to the society.

To the people of the Ughelli oil communities, a disease must have a cause(s) likened to the 'germ theory' of the orthodox medical practice, but different in diagnosis and treatment. The cause of disease is first understood in the context of either the patient's violation of the norms of the society, or had committed an abomination that angered the ancestors. The patient is believed to have been therefore, struck down, or bewitched by the ancestors/ ancestresses. For instance, if a man makes love to his wife or any other woman in the bush regarded as where domestic animals like goat, dog and fouls are not reared, the ethical divinity or *Erivwin* strikes him down with an illness. A woman under menstruation is traditionally forbidden from going to the farm. If she violates, the ethical divinity strikes her down.

These taboos are seen as possible causes of virtually all ailments; in the area. But the actual diagnosis starts with an enquiry into the divinities, ancestors and spirits to first establish its actual cause(s). Should any of the ancestors be found to be angered by the patient, then he/she must appease the ancestor through sacrifice before the real treatment of the disease begins. For instance, if a man made love to his brother's wife, and his ailment has been traced to the annoyance of the ethical spirit, it must first be propitiated and thereafter, the brothers reconciled, before the treatment of patient. Herein lies the holism in the traditional medical practice of the area as opposed to the orthodox medical system.

Although herbs are used in the treatment of patients, their potency is traditionally rooted in the invocation of the power of the divine forces. Rituals are very instrumental to the success of this system of traditional medical practice. Turner succinctly summarizes the role of ritual in all this follows:

> The symbols of ritual are, so to speak, 'storage units' into which are packed the maximum amount of information. They can also be regarded as multifaceted memories, each facet corresponding to specific chister of values, norms, beliefs, sentiments, social roles and relationships within the total cultural system of the community performing the ritual.[9]

The crucial point, as Okpako noted, is that:

> Since rituals performed in pursuit of the restoration of health are usually public affairs, then the basic concepts of the causation of disease and its prevention, death and other misfortunes

> are generally understood by adult members of the traditional community. That is why a diviner or traditional healer has no difficulty in convincing his patients or his relatives that the illness has been caused by offended ancestors or witchcraft and he has the knowledge too, of how to cure it.[10]

The role of the diviner or traditional healer in the traditional medical systems goes beyond the scope of medicine to include the upholding of moral and spiritual values among social groups, reconciling members of the society, and in the process, promotes harmonious relationships between them and the spiritual world. Thus, the 'holistic' nature of traditional medical systems stems from the fact that an offender first confesses his/her misdeed after the diviner has established the cause; thereafter, he/she is reconciled with the aggrieved. Confession is thus an intricate part of the propitiation of the angered ancestors/ancestresses. Unless that is done, rituals do not have any positive effect on the sick. This aspect is the spiritual healing and the most important. The ritual proper is done at the final stage, which involves administering of medical herbs either orally or internally or both, to enable the body to regain its natural anti-bodies that were weakened due to the illness.

Preventive medicine is also a significant aspect of the people's traditional medical system. A foreknowledge of herbs and its application to the health management of the community is crucial in this regard. The area is rich in flora and funa with medical values and the people also have a widespread herbal knowledge. There are cults of traditional medicine whose major occupation is to prepare and administer these drugs. They provide necessary training for young men and women who indicate interest in the profession. The period of training might span fifteen years particularly if a lot of spiritual training is involved. Some illustration of preventive medicine is important at this point.

The water lettuce (pistia stratiotes) locally called in the area as *Itetebe*, is an aquatic weed. It is specifically used to control birth, although it has other medical values. The method of preparation is as follows: grind the herb and thereafter, mix with native chalk (*orhen*); then tie the concoction in modules with a piece of new cloth. While the medicine-man prepares the concoction, he recites incantations and invokes the power of the divine forces. Finally, the medicine is hung around the waist of the women whose pregnancy is threatened by abortion. The bleeding stops and the ritual is performed again before delivery. Another illustration ends this section.

The *Oghriki* tree and the shell of the tropical snail, *Isekpe*, are used for child spacing and for controlling abortion as well. The preparation: the empty shell of the snail is used to scrap off the bark of the tree seven times. Whatever quantity is collected inside is then burnt together with the shell. Thereafter, it is made into powder and taken internally by the women. As usual, the process is done with incantations. For a bleeding expectant woman, the abortion stops immediately while for a woman who needs spacing, she cannot become pregnant until the ritual is repeated. The same ritual is repeated for the pregnant women in the ninth month to enable her deliver the baby. Unfortunately, the basis of the traditional religion and medical systems of the people were all distorted and destroyed by British rule and later re-oriented to cater for the forces of Western imperialism not the least, Shell. Let us now turn to the process of destruction.

7.3 Colonial Period

As noted, the major objective of British rule in Nigeria was specific: to extract raw materials locally and bring in manufactured goods from the metropolitan factories. The instruments that were put in place to achieve its goal included the colonial state, monetization, taxation, the establishment of monopolies, education and missionary groups. However, for the purpose of this chapter, only those modalities of exploitation that more than any other, destroyed the cultural environment, are addressed. To that extent, the colonial state, education and christian missionary groups are discussed, others having been treated reasonably well in Chapter Four of this study. However, their intricate linkages with other agents of underdevelopment will be referred to when necessary.

7.3.1 The Colonial State

The British created the colonial state specifically to facilitate the capitalist exploitation of the Niger Delta in particular and Nigeria generally. To be able to play this role well, it was entrusted with the monopoly of coercion to enforce colonial legislations so that 'law and order' was maintained to permit colonial capitalism to thrive and expand.

Culture dies hard because it is difficult for people to give up easily their native ways of doing things. The cowries for example served two basic purposes among the people of the Ughelli oil areas. First, as a medium of exchange, which also could hinder the expansion of the British monetary system if not annihilated. Second, as a language too used by diviners in virtually all traditional religious practices in the area. The cowrie is in also used during burials. Colonial capitalism was not ready to co-exist with such cultural practices. In 1904, the Anti-Cowries Act was enacted to annihilate it not only as a pre-colonial commodity currency, but also, make it unlawful for anybody to use it for other purposes such as burying the dead. The colonial state saw to the successful implementation of the Act. With such prohibition, the native medical system and traditional religion of the people, for the first time, began to retrogress particularly as the scarcity of cowries and fear of incurring the wrath of the state, hindered the role of the traditional medicinemen in particular, and the entire religious practice of the people generally. Efforts made by the divine healers to improvise with the shells of the 'native mango' have again been frustrated with the advent of Shell whose degradation of the environment led the extinction of the crop in the area. Perhaps, education will shed more light on the damage of the primordial cultural traditions of the people.

7.3.2 Education

In the beginning, the colonial state did not want to introduce formal education for fear of creating an educated and conscious class that would understand its politics and economics of exploitation, thereafter, seek for its overthrow. But it hesitantly did so to reduce the prohibitive cost of administering the colony with expatriate labour. Colonial capitalism could hardly have flourished without introducing an ideology that made the people to consider their traditional beliefs as primitive. Colonial education was used to achieve that objective through the indoctrination of the people with British values; values that were largely rooted in British capitalism.

The first and the only secondary school established in the area by the colonial state was Government College, Ughelli in 1945. Admission into the school was largely restricted to children of those who served the colonial regime, as well as wards of powerful chiefs, merchants and distributors to the colonial trading companies who also facilitated British capitalist exploitation of the area in particular and the country generally.[11]

As to be expected, it was the educated elite, inevitably created by the colonial state, that led the anti-colonial government campaign in the early 1950s, dug its grave and finally buried it in 1960 when Nigeria became politically independent. Whether the post-colonial state even brought development to the area when crude oil was discovered, is an issue discussed in the section on the activities of Shell. The role of colonial education in the erosion of the culture and traditional values of the people went hand-in-hand with the activities of the christian missionaries.

7.3.3 Christian Missionaries

As noted earlier, effective missionary activities in the area actually took root after 1900. The missionary groups adopted a two-pronged approach in what Eme Ekekwe described as 'their civilizing mission of imperialism'[12]. First, they established missionary schools at both primary and secondary levels and admission was limited to the children of converts. The theme of the christian message was 'peace and brotherhood'; and indigenes who were disposed to it, had most of then serving in the colonial regime. The Bible was the sacred book and its scriptures were taught in their schools and churches. For instance, the various divinities were observed as devils and the people were forced to stop the taboos prescribed for them.

The second approach was the establishment of mission hospitals. Admission of students for their nursing programme was restricted to converts. Furthermore, the missionary doctors played two roles: (i) they preached for the immediate discontinuation of the traditional medical systems; and (ii) trained nurses in the use of western medicine. In the churches, sermons by the missionary priests preached against worshipping idols, the revered gods or divinities of the local people. In fact, the activities of the christian missionaries, especially the Anglican and Catholic missions that dominated the area, complemented the role of the state. Ekekwe, again aptly summarizes them as follows:

> The missionary groups and the state also complemented each other, since the mission schools provided the state with some qualified manpower while the state provided the missionaries with protection. In this imperialist mission, both Maxim guns and the Bible found mutual support and benefit.[13]

Let us now discuss the consequences of the operations of Shell on the cultural environment of the area having explained the role of British rule in setting the stage.

7.4 The Impact of Shell's Activities on the Culture and Health of the Oil Communities

7.4.1 The Political Economy of the NNPC/Shell Joint Venture

As explained earlier, it is important to begin the discussion with an explanation of the Joint Venture Agreement under which Shell and the state have been producing crude oil since 1973. It is noted here that the NNPC is the executive organ of the state in the petroleum industry and the corporation, as used here, implies the state.

The basic questions to ask about the joint ventures arrangement are: Does the NNPC really operate a joint venture agreement with Shell? If yes, does it actually produce crude oil for the partners? Is the arrangement a 'participatory share' or a proper joint venture? Is the participatory arrangement significantly different from technical cooperation? Are the oil companies particularly Shell not embracing this kind of economic nationalism by the Nigerian state both as a shield from direct nationalization and a way of reducing risks especially with the rise in anti-oil companies' feeling in the Niger Delta? The significance of these and other related questions to the discussion is that, they touch on the heart of the issue: the hollowness of the entire arrangement and its implication for the pillage of the host of areas. A brief political economy of the joint venture will enable us capture its essence for the Ughelli oil areas.

As noted, the state, through the then Nigerian National oil corporation (NNOC) had, in 1973, and later, the Nigerian National Petroleum Corporation (NNPC) signed what is erroneously believed by many scholars of Nigerian economy to be a joint venture agreement, based on production-sharing, with shell. A joint venture proper as opposed to the NNPC/Shell type, is a type of investment that involves equity participation. However, the agreement the NNPC had signed with Shell is not, as the company is still 100 percent owned by the its parent company, the Royal Dutch/Shell Group in Europe. This explains why its shares are not traded in the Nigerian Stock Market but only in capital markets in Europe and the USA. The NNPC/Shell joint venture is participatory in nature, and no less than a type of technical cooperation in which Shell is the technical partner bearing the entire cost of exploration and production, while the state provides the needed local political climate conducive for oil exploration and production. For instance, the participatory arrangement stipulates the ratio of shares that the partners hold in the partnership and that each partner is expected to contribute in the form of both funds and human capital in line with the ratio of 80:20, NNPC/Shell. This is in theory. In practice, the acreages already awarded to Shell before 1973 are however, not affected by the new arrangement, meaning that it is only in post 1973 acreages that are affected. This is all a contradiction in the way the arrangement is formed.

Since 1973, the state has hardly met its part of the bargain both in funds and manpower. Rather, it has conceded its supposed position as 'senior partner' to Shell based on an understanding that shifted the entire cost of oil exploration and exploitation to the company. Shell accepted this on the ground that the cost of exploration and production of crude oil is calculated in crude oil and its equivalent subtracted from the total output of crude oil produced. The remainder is then shared in the agreed ratio. All further gave Shell an upper hand in the deal, leaving the state weaker.

The major implication of such an agreement for the social cultural lives of the people in the Ughelli oil producing communities is that, Shell continues with its reckless operation knowing fully well that it cannot be held liable for pollution. The same holds for the payment of compensation in which the state inevitably bears the greater share of the cost, though it never paid. The lopsidedness of the arrangement has made the stoppage of flaring of gas, among other sources through which Shell degrades the environment impossible. Not even the Gas Reinjection Decree of 1979 will make Shell to comply. Rather, the company is exploiting the weakness of the arrangement to its advantage. On account of its material interests, the state has not taken any decisive sanctions against Shell Instead, they both hide under the cloak of such an exploitative arrangement to wreck havoc on the cultural environment of the Ughelli oil area among others. Their activities and consequences are examined below.

7.4.2 Exploration

As explained earlier, apart from the Shell's lease-hold on its concessional blocks, virtually all other stages involved in finding, mining, transporting, and storing crude oil are contracted out to the 'independent' or the oil servicing companies (OSCs). This does not account for why Shell is unable to adequately monitor their activities. Rather, it, like the OSCs, is not interested in protecting the environment of a minefield. Furthermore, given the exploitative relationship between Shell and the OSCs, the latter is under no obligation to the host communities to be cautious in their operations. Much more important to note in understanding the recklessness of the OSCs, is the nature of the agreement they have with Shell. Because Shell has a post of OSCs from which it hires, there is keen competition between the servicing companies. So, the ultimate purpose for OSC of this stage, is to find and win crude oil, irrespective of the consequences of their activities for the environment and culture of the people. Let us illustrate with seismic operations.

Seismograph Services Limited, a foreign oil servicing company, has been Shell's major exploration company operating in the area in the past 40 years. Its mandate is to provide seismic data bank on which other oil exploiting contractor companies to Shell will rely to find oil in the area.

The degradation of the cultural environment of the area begins with the operations of the company. The initial step is to delineate an area and thereafter, a party of not less than 400 tool pushers with about five seismic engineers then starts to work. Within the area, over 10 parties will start work simultaneously. Which means over 4,000 Seismic will be ravaging an area in a single operation. After delineating an area for operation, the next step is the clearing of pathways on which the tramps are laid usually at intervals of not more than 200 meters. Pipes are drilled into the crust with the necessary explosives planted at strategic places within the areas. The system is connected to a sensor that records the information of the nature of the synclines that bear oil after the explosives have been detonated. This technical details are necessary in order to understand in full, how they damage the environment. These activities are carried out not only in the swampy areas and forests but even within towns and villages. Shell and Seismograph Services Nigeria Limited operate regardless of the sacredness of sites and threats to the safety of the inhabitants of the host area as most of their pathways and tramps run across notable shrines. So, they willfully undermine the privacy of the inhabitants of the area whose traditional bathrooms and toilets, usually the uncleared bushes at the back of their houses, are assaulted in the process. As explained earlier, most of the divinities in the area have their shrines located either in the forest and swampy areas, or even at the banks of the rivers and creeks. Associated with each divinity, are a lot of taboos, such as the prohibition of the shooting of gun or detonating explosives unless after the priests/priestesses of these gods have consulted them and approval secured. In some cases people are requested to pull off clothes before crossing areas where they are located. Non-indigenes and foreigners are not allowed to enter such sacred sites too. Shell and its OSCs do not recognize these cultural values. Hence their workers violate them with impunity. Some of the clans and their totemic animals are already shown ins Table 7.1, are incidentally also the oil-producing areas. The seismic crew of both Shell and its OSC have repeatedly criss-crossed these shrines and violated all the taboos observed for the deities. In fact, there was hardly any community whose traditional places of worship have not been invaded by the oil workers at the time this study was conducted. Some illustration is idea here.

What the people of Ughelli the author spoke to, described as unprecedent took place in February 1993. The shrine of the *Umanoku* was trespassed by Shell and its contractor companies, bringing into the open, the hitherto mystical beliefs of the power of the divinity. Except the priests and offenders who come made to appease the god, its shrine is not a tourist centre. No other persons, much less strangers, are allowed to get close to it. All this was violated repeatedly when the oil workers most of whom did not hail from the area assaulted the bush covering the shrine while carrying out seismic operation there. That is not all. In that same month, one of the oldest sacred sites in Ughelli which is visited only during the coronation of a new king had seismic cables laid across it without the knowledge of the priest. The experiences of other communities are not different. For instance, within the Agbarha oil/gas field, the exploration activities of Shell had jeopardized their religious and native medical systems. Due to frequent seismic surveys in the area, the inhabitants of Awhire, Omavowe among other villages along the Kokori/Orogun/Agbarha angle, have had their shrines and sacred sites not only criss-crossed with tramps, but bulldozed An act believed to compel the ancestral and marine spirits of these shrines to depart the area, with, In turn, made many of them to feel unprotected.

7.4.3 Production/Gas Flaring

No other activities of Shell have seriously destroyed the culture of the people of the area than oil production including gas flaring. The reasons are not far to seek. A lot of stages are involved after exploration, before an oil well can be said to be producing. The technical details may not be necessary but it is important to note that over five stages with the services of not less than five different oil servicing companies are required. And at each stage, the traditional religion and native medical systems of the inhabitants of the area are distorted and/or destroyed in one way or the other. Let us describe them beginning with drilling, which is the next stage after exploration.

The drilling stage, like exploration, is contracted out to the oil servicing companies. But since drilling is associated with many other technical services such as well logging, they are all discussed together. Shell's major OSCs at this stage are foreign. Bariod Nigeria Limited, Forasol Drilling West Africa Limited, Baker Nigeria Limited, Dowell Schlumberger Nigeria Limited, Flopetrol Nigeria Limited, Anadril Nigeria Limited, Ecodrill Nigeria Limited, Otis Nigeria Limited, Saipem Nigeria Limited, Sedco Forex Nigeria Limited and Schlumberger Nigeria Limited are among them. Details on this, is contained in the appendix. Since a separate section treats the activities of the oil workers, it suffices to note here that as in the exploration stage, the workers of these companies still wreck havoc on the native religions and medical practices of the people.

The activities of the drilling and its associated companies are land-based including the swampy areas. They cut the well formations, cushion them and use drilling muds. Cement slurry and other acid stimulations are used also. The wastes from all these chemical are harmful to the flora and fauna of the environment including even human beings. Yet, they are not treated before dumping them on the environment since the area is regarded as a minefield. At times, pits are dug and into which the wastes are dumped. Pits that are not cased at the bottom and top, the wastes are buried, and therefore, overflow or spill into other areas mostly in the rainy seasons. Because most of the chemicals used by the contractor oil servicing companies to Shell are water-based, the depth and speed of pollution are so fast particularly in the swampy areas that within an average of twenty-four hours, over five communities, are all polluted. Notable water-based chemicals used by Shell for drilling are Bentonite whose wastes cause bronchial problem if discharged into the environment unlocated. Others are soda ash, EP-20 with equally dangerous wastes. A list of these chemicals was already continued in Table 6.2.

One of the immediate effects of these pollutants particularly on the fauna is the killing of the totemic animals of the inhabitants of the affected communities, both during and after oil production. Those that are able to escape, especially those in the reptile family, such as the python, alligator, crocodile etc., crawl into the farmlands and to the villages. By tradition, the Usiefurun people, for instance, are forbidden from seeing the python around their homes *except during their traditional festivals*. For, its presence signifies danger. The same is true for the antelope among the people of Ughelli. Due to the constant pollution of the swampy areas where most of these revered animals live, the people have been forced not only to be seeing their totems constantly but even to be discovering dead ones in their farmlands and in the swamps. Prolonged dry seasons in the area are believed by the people to be caused by the angered ancestors and divinities. Appeasing the angered divinities does not have any significance, as no sooner has one ritual been performed than another abomination committed. In fact, this aspect of the pollution of the area is widespread, implying that every clan in the Ughelli oil areas has at one time or other, been exposed to artificial life and hardship as a result of the operations of Shell.

The effects of the water-based chemicals on the flora of the area are equally negative particularly on the water weeds including those with herbal values. However, the magnitude of the destruction of the vegetal cover is highest under another form of pollution: flaring of gas. The joint venture under which both Shell and the state produce crude oil, is a hoax. Shell's continuous flaring of gas in the area since 1958, and even after the enactment of Gas Reinjection Decree of 1979, is a clear manifestation of its hollowness. For instance, Shell willingly flouted the 1979 Decree because of the state's unwillingness to obey its own law. The position of Shell is specific; that is, it can bring into the country, a sister company from Europe whose specialty is in gas reinjection, insisting that the cost will have to be charged to the joint venture. Allowing Shell to have its way means that the state will bear 80 percent of the cost of gas reinjection. The state objected to the position of Shell, arguing that Shell, as the technical operator, has to protect the oil areas by redirecting gas into the earth's crust. But the option of fine which the state took, shows that if, too, is not serious with the protection of the environment. It is the oil-producing areas, this time, the Ughelli oil communities, that bore the brunt of the such form of pollution.

All through the Eruomukowharie, Olomu, Afiesere, Erwreni and Orogun oil-producing areas, Shell has never stopped flaring gas for the past years it has begun operation there. There is an indication that the trend will continue for many years to come. Apart from the local deliveries made to NNPC from which Delta Steel Company, the National Electric Power Authority, Sapele, Shell's direct connection to its staff quarters at Warri and Port Harcourt, all accounting for about only 25 percent, the remaining 75 percent of the total gas produced by the company is flared. In 1987, for instance, of the 8,490,937,646 cubic meter of gas were produced by Shell, 65 percent, or 5,452,758,411 cubic meter, were flared. 2,157,439,299 cubic meter, representing 35 percent was used locally by Shell, other private and public companies[14]. Between 1989 and 1992, when popular protests wanted the then Ibrahim Babangida's regime out of power, there was a hill in the companies consuming natural gas. As a result, local consumption of gas fell to less than 10 percent. This meant that Shell had to flare over 80 percent of its gas in the oil locations since all NNPC's gas pumping stations located alongside the company's oil' gas field were vandalized and work stopped. For reverting to the pre-joint venture era under then Abacha junta, the state signalled its willingness to invest in exploration and production in partnership with Shell and other oil companies. The regime of Obasanjo seems not ready to address the issue decisively. All has made Shell freer to flare gas willy-nilly.

The effects of gas flaring on the culture and health of the people in the Ughelli oil area can be understood in two basic contexts. First, its effects on the herbal plants. Apart from the knowledge of the medicinal value of plants among the people as earlier explained, the area is equally rich in medical herbs. The forest serves as the natural pharmaceutical gardens from which the native medical doctors collect most of their herb. But these herbs have almost been depleted as a result of the heat generated from the continuous flaring of gas by Shell. They are depleting so fast because most of the flare sites are diverted away from where the oil well head is located to prevent fire outbreak in the locations in the forests, farmlands and swamps. The most common feature in such areas is a 'derived Savannah' in a forest region. For instance, the two flare sites in the Kokori/Orogun oil and gas field located at Erhoike and Erhobaro that used to be marshy areas and forest respectively, and rich in herbal plants have been turned into semi-savannah. In Otu-Jeremi, which is less than 200 meters from the Utogogu gas plant, the heat radiating from the gas flare, is depleting the herbal plants even in the compounds of the inhabitants. This is compounded by the intense heat of about 2000°C being emitted from the gas plant into the host area. One of the notable effects of all this, to that the 'native plantain' whose leaves are used in burial rites and for medical purposes too, have gradually become extinct. Table 7.2 presents some of the herbal plants being depleted in the area by flaring of gas.

Table 7.2: Selected Depleting Herbal Plants in the Ughelli Areas Due to Shell's Environmental Pollution

Community/family	Botanical	Traditional Medical Use
Water lettuce (aquatic weed)	*Pistia Stratiotes* (Linn)	Breast feeding, birth control and tapeworm
CYPERACEAE (aquatic weed)	*Pycrus Lanceolatus* (Poir)	Adrenal gland stimulant, blood radiation and aura
Resurrection plant (aquatic weed)	*Bryophyllium Pinnatum* (Linn)	Cough, sores and hypertension
Common Club Moss	Lycepedium clavatun	Cramps, rheumatism, and headache

Source: Author's fieldwork, September 1993.

As a result of the flaring of gas, one of the bases of the native health-care system of the area has been destroyed. Scarcity of these herbal plants in the various oil communities as shown in

Table 7.2 above, makes the practice of traditional health all the more difficult. The state-owned general hospital in the area, built in 1964 was, in most cases, lacked drugs. What is more, gas flaring has had an adverse impact on the local concept of time, and on the totemic animals. Time is as crucial as the herbs in the preparation of the native drugs and concoctions. Most of the divine healers do not only communicate with the divine forces but also perform their rituals at night. Shell's flaring of gas in the area for both day and night, and in the past 40 years, has impacted negatively on the herbalists traditional concept of night since the night is lit up making it look like day. Certain sacrifices that are done at night now take place in the gas-lit night, causing the divine forces to reject such sacrifices. So, traditional healers in the area have been practicing their professions with greater difficulty as diseases become resistant because of the constant violation of the taboos associated with the gods by the Shell and the state.

One important effect of the flaring of gas on the totemic animals is that the heat generated has forced them to migrate to areas outside of the flare site. Since there is hardly any community where Shell is not flaring gas, the general trend is that most of the totemic animals as shown in Table 7.1, have been forced to flee the area. In effect, the people are traditionally believed to be living without guidance from their ancestors. For instance, the presence of these revered animals during rituals and sacrifices that precede traditional festivals in the area, is symbolic of acceptance by the divinities. Because their places of abode have been invaded by the oil companies, it is difficult for the priests/priestesses and native medicinemen to communicate with them or calling them during rituals. The sound from the detonation of dynamites by the seismic companies has hastened their flight.

7.4.4 Transportation

Once an oil well begins to produce, Shell transports crude oil and gas through pipelines first to the flow station, then the crude to the terminals. The NNPC also gets it deliveries of natural gas from the company and then supplies its consumers through pipelines. So far, pipeline is about the cheapest and safest means of moving crude oil and natural gas. Shell's only Quality Control Center to which all its flow stations are linked in the western division, is located at Ughelli. Before the Forcados terminal came on stream in 1971, it was from the Center that the crude after being treated, was pumped to the Bonny terminal for export. With the Forcados terminal on stream, and mix oil wells sunk in the area, Shell built up more network of oil pipelines from its various flow stations scattered all over Delta State. All is with consequences for the culture of the people.

As explained earlier, Shrines of the divinities are sacred sites, which by the tradition of the people should not be moved. Shell, has violated and continue to violate this aspect of their tradition with impunity. Cases abound in the area whereby villagers have been forced to shamefully pick up the pieces of structures made as their gods after bulldozing shrines that fall on the company's right of way for oil gas pipelines. Over 40 villages in the Orogun and Agbarha region of Shell locations have suffered from the disgraceful act. The Igbe religious groups in the area have, for want of sacred sites, resorted to using the deep burrow pits left over by Shell's contractor construction companies like Niger Cat, as shrines.

Not only that. The pipelines are laid bare on the ground and across the farmlands, sacred bushes, and compounds of the inhabitants of the area. For instance, Shell's oil/gas pipelines are laid bare on the ground right from the Kokori/Orogun oil and gas field, through the Eruomukowharie gas field to that of Umolo. Shell has a network of 1000 kilometers of oil/gas pipeline in the area. Quite apart from posing danger to the lives of the people, the *Oghene-egodo*, among others, that is, god of the compound, is believed to be angered by the imposition of the 'foreign culture' on its domain. The power of the god of the compound is believed by the people to have been gradually eroded, thereby breaking the very fabric that traditionally held the family unit together. Since many of the activities of the oil companies are being carried out by workers who do not hail from the host oil communities, they do not hesitate to violate the culture of the area.

7.4.5 The Oil Workers

One of the effects of the complete integration of the Niger Delta into the British capitalist fold by 1900, was the creation of a pool of unskilled urban labourers who, in an attempt to survival, must sell their labour power at whatever wage. And it was from this pool that Shell and its contractor companies first recruited most of their casual workers. Furthermore, the globalization of the oil industry made the expatriate staff of Shell and its contractor companies a blend of various nationalities: France, Dutch, British, Swedish, American and African countries. In essence, Shell's workers constitute a 'new community', where the only culture is that of exploiting, producing and transporting crude oil cheaply. The introduction of christianity as earlier explained, made the 'no-culture-attitude' of the oil workers to take root. All has come to explain why the oil workers, local and foreign, violate the cultural beliefs of the people in the course of finding and producing crude oil. For instance, most of the totemic animals already shown in Table 7.1, are delicacies to the oil workers. And they kill them as they search for crude oil irrespective of the fact that they are revered.

Camps are usually built by the OSCs close to areas where they are prospecting crude oil. The local workers and a handful of the expatriates retire to these camps at the end of the day's job. By the tradition of the local people, these camps are bushes because domestic animals, especially goats and chickens, are not reared around them. In this context, it is therefore, an abomination to make love to woman there. But these camps are the make-shift homes of the oil workers. It is in these places that women are lured into for sex by the workers. Some of them even sneaked into the villages at night in search of women who were brought into these camps for sex. By so doing, they heightened promiscuity among women in the area. One of the consequences is that there is hardly any village in the Ughelli oil areas where 'fatherless' children are not found. Mothers of these children were unable to trace their fathers because as soon as the oil workers move to another location, they never return there. Apart from inflicting social cost on the women, the society in general has been traditionally 'polluted', because 'bastards' are taboos in these communities. Cases of rape and adultery have become so common that sacrifices offered to the ethical divinity, are no longer cleansing the land of the immoral acts of the oil workers.

7.4.6 Oil Spillage

As noted earlier, both the state and Shell do not recognize that there is an oil community in the Niger Delta. Rather, the areas where crude oil is explored and produced, are minefield. In effect, the inhabitants of the oil communities are squatters in their ancestral homes, without right to complain about oil spillage and pollution of their environment including that of their culture. It, therefore, does not make any sense to both Shell and the state when the people of the Ughelli communities, like other oil bearing areas, protest against the degradation of their environment because oil spillage is regarded by the company and the state as one of the natural hazards to be expected in a minefield. It is with such understanding that both Shell and the state operate in the Ughelli oil areas among other oil-rich parts of the Niger Delta.

Studies have shown that crude oil can be caused to spill as a result one or a combination of the following: blowout, equipment failure, maintenance error, engineering error, third party sabotage, sand-cut, accident and even by unknown factors. Spillages in the area have had over 80 percent of their causes traced to equipment failure and maintenance errors, particularly burst oil pipes. Until an oil pipe bursts, Shell for instance, does not deem it necessary to carry out cathodic maintenance. Since 1976, oil spillage resulting from broken pipes has been an annual occurrence with an average of 8,000 barrels of crude oil lost to the environment in each spill. The most recent ones were at Otu-Jeremi in April 1994 where about 100,000 barrels of crude oil were lost; another at Ovwodokpokpo in July 1993, in which an estimated 150,000 barrels were all lost. At the Otu-Jeremi incidence in February 1994, about 200,000 barrels were spilled and all lost to the environment[15]. Within the period under investigation, that is 1976-1994, there has been no spillage caused by blowout. Although Shell and the State have complained of sabotage, they hardly support their claims with evidence; rather, the accusations are advanced to avoid paying compensation to the affected communities.

The frequency of oil spillage in the area is damaging and destroying the flora and fauna of the immediate environment. Much more unfortunate is the levity with which both the state and Shell handle the mop up exercise and environmental impact assessment. To begin with, the state relies on Shell for the figures on the barrels of crude oil spilled and reports on its causes as well. It is therefore unrealistic to expect the company to give the accurate figures and prepare a report that indicts its own activities. The state on its part, does not brace up the challenges by carrying out independent investigation and research. Rather, it only accuses the people of sabotage and mobilizes its armed forces to intimidate and repress them. So, after any spillage, no attempts are made to clean up the oil, or to set up a commission to assess its impact on the socio-cultural life of the people. The people have just one choice: they clean up the oil sleaks through communal labour. But before doing that, the spilled crude oil has percolated into the soil with the attendant damage to the surface and underground water, among other dangers.

7.5 Reactions of the Inhabitants of the Oil Communities to Exploitations and Responses of Shell and the Nigerian State

Like the colonial trading companies, Shell's main concern in the Ughelli areas is to find and produce crude oil profitably. It did this with all the negative impacts on the culture of the people of its host communities. Perhaps, the company hopes that the inhabitants of the area will ever remain ignorant of the consequences of oil spillage and flaring of gas among other forms of pollution. But that is not to be as almost every oil community is gradually becoming aware of the dangers that oil exploration and production pose for its people, their environment and culture. Though there has been pockets of protests in the area, the launching of the Ogoni Bill of Right in 1981 by the MOSOP had a spill-over effect on other oil communities. As a result, the people of the Ughelli oil areas began to intensify demand from Shell for compensation for the crude oil it has extracted from their land, insisting that it respects their tradition while prospecting and exploiting crude oil there. They also took legal actions against Shell. Protests have been violent as well. It became clear to Shell that the inhabitants of the Ughelli oil areas are poised to halt its operation should the trend of degrading their environment continue unabated. Shell's earlier argument that it is a junior partner to the state, and just a technical operator of the NNPC/Shell joint venture, is not helpful because anti-Shell feelings are mounting in the area and other host oil-producing communities across the Niger Delta.

The state, like its colonial predecessor, was initially only interested in the proceeds from oil taxes, royalties and exports; and, therefore, did not concern itself with the safety of the environment of the oil communities. But as Shell's operations are regularly disrupted and at times, halted, it became more repressive. In fact, the state has deployed its armed forces to quell any threat to oil production, storage and transportation; and even stationed some of the units in the area to guard the oil installation against vandalization. This is all the more so because crude oil has, since the turn of the 1970s, become the mainstay of the country's economy. From all appearances, both Shell and the state are under attack from the people of the Ughelli oil areas. Let us now discuss the reactions of the people.

7.5.1 Legal Actions

The first reaction the people took against Shell for its prolonged degradation of their environment was to seek redress in the court of law. Their main argument is that, as the custodian and interpreter of the laws of the land, the court should be an impartial arbiter and their last hope. But the court is part and parcel of the state structure, making it difficult to rule against the state. This is all the more so since crude oil accounts for over 90 percent of the country's export earnings. Rather than remain an impartial arbiter, charges of environmental pollution were politicized by the court. Because the protesting inhabitants of the area are poor, they lack the funds to hire experts in ecology who would provide proofs of environmental pollution. For instance, most of the complaints the people took to court bothered on negligence for which they could hardly show evidence. Let us illustrate this with the case of Mr. A. Atubin versus Shell.

In 1973, A. Atubin, leading members of oil-producing communities under Olomu clan, sued Shell at the Ughelli High court for causing crude oil, gas and other dangerous chemicals to escape from the oil pipelines under its control. And as a result of the spillage, fishes and herbal water weeds were killed, and their rivers and creeks were polluted. Their sacred places of worship located on the swampy areas were not spared as the vegetal cover began to wither due to the pollution. In particular, oil bean trees, revered plant of the people, were badly hit. They demanded for a compensation of N5 million. The trial judge held that the plaintiff must establish the linkage between the breach of duty and harm done in spite of the overwhelming evidence of spilled crude still floating in the swamps of the area. Because the plaintiff and his lawyer could not do that to and did not bother to hire Botanists and Biochemists to carry out a detail study of the area in order to establish the type and magnitude of harm done to the environment, the judge dismissed the case on technical ground that the plaintiff was unable to provide evidence to show that the negligence of the defendant (Shell) was the direct cause of the damage. Their charge of public nuisance against Shell was also struck out for want of evidence[16]. It is important to note that all the oil communities in the Ughelli area have waged futile legal battles with Shell for negligence and public nuisance since 1973. All the same, each community is becoming more aware of the methods involved in seeking redress against Shell and the state. It has also become clearer to them that the fight against environmental degradation, including the protection of their culture, is capital-intensive and that no community can win it without experts of the Biological Sciences. It was partly to achieve this aim that the Urhobo Study Group, a body of Urhobo intellectuals from all disciplines, was set up in 1992. As a relatively new body, what is preoccupied with at the moment, is to assess the level of pollution of the oil areas in Urhobo land.

7.5.2 Popular Protests

Our focus here is on the protest by the people against the underdevelopment of their culture by Shell and the state. It is important to state too, that the protests in the land are holistic in the sense that all aspects of the life of the people – social, cultural, economic and even political – are affected by the capitalist development promoted by Shell and the state in the area.

Popular protests in the area have taken two major forms: peaceful and violent. First, the peaceful protests. Being conscious of the law, organizers of peaceful demonstrations express their demands whenever officials of the government such as governors, ministers, and heads of state visit the area. When then President Shehu Shagari visited the Ughelli area in 1982, most of the placards read: 'Shell must go!', 'Protect our Environment!'. Even then, the protesting groups were shielded away from the President. It is not that if he saw them, something concrete would have been done. For, past experience had shown that, political promises were usually made. The situation has been different under the military regimes which did not even allow any form of protest, no matter how peaceful. Since Buhari junta in 1983, there has been no such peaceful demonstration in the area because the junta prohibited it. For instance, when the then military dictator, General Ibrahim Babangida, came to commission the Utorogu Gas Plant in 1991, an area under the Ughelli oil areas, attempts by the people to demonstrate peacefully were nipped in the bud by security officials. Again in 1992 when Babangida paid an official visit to Delta State and came to Agbarha-Otor and Ughelli, no protest of any kind was allowed. As it stands, peaceful demonstration will not bring the desired impact.

Violent protests, on the other hand must be properly planned and executed in order to achieve their objectives; which is to disrupt oil production, transportation and storage. Although a leader similar to 'Ken Saro-Wiwa' is yet to emerge to direct protest in the Ughelli area, violent demonstrations have made some impact on the disruption of the activities of Shell. They have taken three main forms, namely, solidarity with other oil areas in the Niger Delta, proxy protests by attacking contractor companies to Shell, and political protest to redress the imbalance in the country. Solidarity protests normally sympathize with other aggrieved oil communities either within the state or in other oil-producing states within the Niger Delta. Geographical contiguity is usually an important factor for any successful sympathy protest. For instance, there is one highway connecting Ughelli areas with other oil communities in Bayelsa and Rivers States. Shell had about 200 oil wells and 8 rigs in the various communities along this route. Once any of the communities embarks on a protest and gathers solidarity support from other areas, the activities of Shell and other oil companies in the area will come to an abrupt end. This is because government vehicles, those belonging to Shell, its contractors and any other oil companies, will be impounded and the oil workers beaten up or held hostage. For instance, in May 1993, the people of Ozoro/Ogoni one of the oil areas in Isoko Local Government, seized the vehicles of seismograph Services Limited. It got support from the people of the Ughelli oil areas who sent warnings to the same company not to visit their area. And for about two months, the company could not carry out seismic surveys in the area.

Protests by proxy are now most common in the area since the people can hardly confront the military might of the state on their own. Proxy protest is paying off because the main targets are the contractor companies to Shell, local or foreign. The wisdom of such actions is that, since Shell remains adamant to their plights, then it must not be allowed to start seismic activities in the area in the first place. In effect, it means that the company cannot discover new oil wells in the land. For instance, in April 1993, the Seismograph Services Limited was prevented by the youths of Orogun from exploring for crude oil in their area. They gave conditions which the company must meet before it could carrying out its work. They are: (i) it has to tar the swampy portion of the road linking the area with Ughelli, (ii) negotiate directly with the owners of the land where oil prospecting is to be done, (iii) respect sacred sites and (iv) warn the oil workers to stay clear of the wives and girls of the people of the host communities. Shell was forced to negotiate with the people before its contractor was allowed into the area. The youth of the community were detailed out to monitor the activities of the company and to ensure that it does not renege on its promises. The women also laid siege on the company's access road to prevent Shell from finding oil in the area. In fact, the women of Ajakota oil area, also overtook Shell's 13 oil wells by barricading the company's access roads in the area for 18 days in 1995.

7.5.3 Sabotage

As noted, over 80 percent of the protests in the Niger Delta area are in Shell's locations. The company always believed that as the pioneer and largest oil producing company in Nigeria, nothing could be done to disrupt its operation because the state can always use its armoured tanks to crush protests when the situation is getting out of hand. The state, on its part, is ever ready to promulgate decrees that protest its most viable accumulative base - the oil sector. But through sabotage, Shell was shaken to its foundation, while the state is caught unaware. For instance, they know when the oil pipelines are not carrying crude oil; they are also informed about the period of maintenance of oil well heads. Such confidential information can only be supplied by an insider; hence the worry of Shell. It is during any of these 'safe' periods that the pipelines are cut so that as soon as production resumes, the spill starts; and it takes Shell about two weeks to detect such faults because the oil-pipes are cut at points that are remote and far into the forest. For instance, in 1992, over 100,000 barrels of crude oil was lost through sabotage at Ovwodokpokpo; and about 4 million barrels lost in the Ughelli area in the last three years.[17] Although the 1975 Anti-Sabotage Decree prescribed the death penalty for those caught cutting oil pipes, among other offences, it has not deterred the people. Since 1980, oil locations of Shell in the area have been under guard by soldiers, but sabotage has deepened.

7.5.4 Popular Movements

In the last ten years, the youths in the area have become more concerned about the general underdevelopment of the area, and in particular, the speed with which the activities of Shell and the state have almost completely destroyed the foundation of their cultural existence among others. They have chosen to check the pillage of their area through the formation of associations under which they will collectively resist Shell and its local allies. So, each community has a youth association. With the increasing awareness of the dangers of oil production in the area, across-the-community-youth associations are beginning to emerge to form what can be regarded as popular youth movements in the area. They may not be as strong as the Movement of the Survival of Ogoni People (MOSOP), but something of that kind is in the making. For instance, it is the youth movement that carries out the proxy protest against Shell, its contractors and even local allies, and the state. The youth movements were also acting as watchdogs on the local allies of Shell in their communities and do not hesitate to support acts, such as the dethronement of the traditional ruler of Uzere, who received gratifications from Shell. Or providing logistics for the women revolt against the Ovie of Effurun in 1992.

The Urhobo Study Group spontaneously created in 1992, can be regarded as a movement of intellectuals of Urhobo stock, whose major objective is to help stem the recklessness of Shell and the state in the area. This objective is to be achieved through research by its members into all facets of the activities of the oil companies and the state in Urhoboland. With facts and figures, the Group believes that the resistance against exploitation by Shell and the state can be a success. For instance, with the assistance of the Group, the impact of any spillage in the community is assessed by the experts, thereby making it easier for the aggrieved people to confront Shell and demand for compensation. The group has assigned some of its members to look into the damage caused to the wetland by Shell in the area, all of which all enable the people to effectively confront both Shell and the state.

7.6 Responses from the State and Shell

The Nigerian state is less of an actor in the production of crude oil but more of a collector of proceeds from Shell among other oil companies. To the state managers therefore, the restive oil communities constitute a major threat to its source of wealth. This explains why it is repressive. On its part, Shell has reluctantly employed direct negotiations with the various communities promising to meet their demands. How much it has been able to do so, will be made clear in this section.

Throughout the 85 days tenure of the then Shonekan led Interim National Government in 1993, all Shell's locations in the area and the NNPC's gas stations were closed down because of the protracted riots and threats to kill oil workers who dared the people. But the Abacha led junta that struck in November of that year, brought temporary peace through repression. However, it dawned on the state that repressive laws and oppression are bringing more losses than revenue. There is need therefore, to use some cosmetic measures to pacify the people. An approach that the Babangida's junta introduced using the OMPADEC to play the carrot and stick politics there. Carrot is usually plum contracts; while the stick is state repression. For instance, the Commission uses the 'carrot' to incite the people against each other with the hope of destroying the bond of unity among them. Until 1992 when the OMPADEC was established, the 1.5 percent which the state claimed to have allocated to the oil producing states from the federation account, was never paid. Even after the creation of the Commission, the money is still not paid. Babangida knew very well that his regime lacked legitimacy not the least because it was a military junta. That the oil areas were in utter neglect, deprivation and underdeveloped was quite known to his government as well. Thus the introduction of 'clientelism' will obviously win a faction of the protesting people of the oil areas to his side. The clients maintain the *status quo*; and to be able to do that effectively, the Commission awarded them contracts most of which were and still usually not executed. The greatest beneficiaries were the traditional rulers, village heads, retired military generals, top civil servants and even the so-called 'newbreed politicians'. The struggle to get access to the Commission became intense as its activities promote rent-seeking. All is to lead to the formation of The Association of Traditional Rulers in the Oil Producing Areas. Association of OMPADEC Contractors among others. It bred hatred among the people as some groups are favoured. For instance, the Okpe people declared that they are not Urhobos. The Ovie of Ughelli also told the Commission that the Ughelli people have nothing in common with the Isokos in spite of the fact that they speak almost the same language. It was all done in a bid to get from the contracts from the Commission. Some illustration with the experience of the Ughelli area is ideal at this point.

The Ughelli oil areas have seven clans, with each of them oil-bearing. Each clan has a head or king as the traditional political head as the case may be. The Commission, given its orientation, awarded contracts only to the pro-government forces made up of people and groups who did not challenge the state and Shell or even the Commission to look into the level of pollution of their environment including the cultural environment. Rather, those who like the state, think that all that the oil areas need are social amenities such as water,hospital etc. in the midst of the decay of the area. So, throughout the land, the beneficiaries of the Commission's contracts were basically traditional rulers and their village heads, spokesmen, retired military officers and civil servants, politicians. For instance, a N1.5 million contract was awarded to one U. Ojefia, a high chief in Orogun, to sink a water borehole for the people. The Ovie of Ughelli had, at various times, reported to have benefitted from the Commission's unspecified contracts to the tune of N20 million. So, too, the Orodje of Okpe, who led his people out of the Urhobo kingdom, got a N30 million portfolio contract. The Dafinone & Company owned by a former Senator in the Second Republic, was reported to have received a N250 million contract to audit the account of the Commission. The Chairmen of the Ughelli Council areas and their councilors also got various unexecuted contracts from the Commission amounting to the tune of N80 million.[18]

However, it was one thing to award contracts but another, and perhaps, more important, to execute it properly. As pro-government elements, the beneficiaries know that the contracts were 'settlements' and that the might of the state could be called upon in case of unrest. At a result, the contracts were either badly executed or not done at all. In Orogun, the chief was sanctioned by the youths for doing a bad job and was compelled to refund the money. He did not pay any money at the time the study was conducted. In every clan, there was the trend of poor quality of job done, or not done at all, thereby increasing the level of resentment of the state and Shell there. With the current financial crisis facing the Commission, the carrots have been exhausted and the state is now using the stick. As noted, the OMPADEC may be replaced by the NDDC if the bill sails through. But that is in assurance that it will be better.

7.6.1 Compensation for Damage

Our analysis in the preceding sections has shown how Shell indiscriminately dumped dangerous untreated wastes on land, into swamps and rivers. These wastes that not only damage the flora and fauna of the area, but also, kill to human beings if the contaminated fish or fruits are eaten; or the dust from the wastes is inhaled. Hear how the Urhobo Oil Communities presented their pathetic plight to the ministerial committee that visited the area:

> ... most rivers in the communities have been confirmed to contained mercury and lead - two dangerous chemicals blacklisted worldwide as causes of human disorder and anaemia ... At Erhoike, gas flaring is felt 300 meters from flare site ... and vegetation 500 meters away are biologically dead.[19]

Until 1973, when it signed joint venture agreement with the state, Shell negotiated directly with the members of the community whenever cases of damages arose. The chiefs and traditional rulers usually represented the people, particularly because they were the custodians of the communal land before the Land Use Decree was promulgated in 1978. That is not to assume that they (communities) had a fair deal because Shell had always preferred the state to act as a go-between. Even the joint venture had created that opportunity for the company to pass back. The Land Use Decree of 1978 therefore, facilitated its exploitation of the oil producing communities. In compiling the list of compensable items as shown in Table 6.5, Shell and the state deliberately excluded the cultural environment so that when it pollutes the swampy areas and kill revered animals and plants, no compensation is paid. Only economic crops and trees are, at times, paid for, with rates as ridiculously low as 5 kobo for a mature bitter leaf, and nothing for the land. To worsen matters, Shell charges such payments to the joint venture account in which it's share is only 20 percent, with the remaining 80 percent borne by the state. In other words, Shell is not really paying compensation.

Finally, Shell's access roads are not social amenities aimed at developing the area. Rather, they facilitated its exploitation.

The company has been compelled by the continual unrest in the area to embark upon some social services like providing bursary allowances to few indigenes of the area in the universities and studying courses such as chemical and mechanical engineering, geophysics, statistics, geology which are relevant to its operations. The allowances are meant to subsidize the students, and the maximum per student per year is only N10,000. The duration of the bursary is four years. However, only 43 indigenes benefitted from it in the area at the time the study was conducted.

7.7 Concluding Remarks

By destroying the sociological basis of the culture and traditional medical systems of the Ughelli oil areas, both Shell and the state have made it pretty tough for the people who live there to experience the kind of development that respects their shared values and norms, which they had enjoyed before the advent of British rule and discovery of crude oil there. Reverting to those old values of the Ughelli people is not particularly an impossible task if the state so desires. In fact, a people without a culture cannot embark on any meaningful development. The state is aware of this. Yet, It has decided to back up Shell in the destruction of the cultural environment of the Ughelli oil areas, among others in the Niger Delta because it profits its custodians. Let us now turn to how its alliance with Shell has heightened land crisis in the rural Isoko oil areas of the Niger Delta.

Notes and References

1. See Julius Ihonvbere and Timothy Shaw, *Towards a Political Economy of Nigeria: Petroleum and Politics at the (Semi-Periphery*, (Aldershot: Avery) 1988; and P.U. Onyige, "The Impact of Mineral Oil Exploitation in the Rural Communities of Nigeria" University of Birmingham: Unpublished Ph.D Thesis, 1979.
2. National Population Census, 1992.
3. Based on the authors field trips to these communities between 1989 and 1994. He also visited Shell's West Divisional Office located at Warri within the same period. The author found out that Shell was reluctant to give details on its oil well in the area basically for fear of reprisal.
4. For details, see E. Taylor, *Primitive Culture*, (New York: J. Murray), 1981; and John S. Mbiti, *Africans Religions and Philosophy*, (London: Heinemann) 1976.
5. See John Mbiti, *African Religious... ibid.*

6. See World Health Organization, *African Traditional Medicine, Series 1, Afro Rep.*, Geneva: WHO) 1976, pp.1-3.
7. See David T. Okpako, "Culture and Development Medicine", in Peter Ekeh and G. Ashiwaju (eds), *Nigeria Since Independence: The First 25 Years, Vol.II*,(Ibadan: Heinemann), 1989.
8. A lot of the first hand information on the subject was obtained by the author from his interviews with the elders, chiefs and traditional rulers in Agharha, Olomu, Orogun, Ughelli clans between 1989 and 1994. Using interpreters, Most of the interviews were conducted in the various dialects of these communities and later translated into English language.
9. See W. Turner, *Primitive Culture... ibid* p.1.
10. See David T. Okpako, "Culture and Development Medicine" in Peter Ekeh and G. Ashiwaju (eds) *Nigeria Since Independence.. op. cit.*, p.223.
11. See Obaro Ikime, *Rivalry in the Niger Delta: Itsekiri-Urhobo Relations and European Presence, 1884-1936... op. cit.*
12. See Eme Ekekwe, *Class and State in Nigeria*, (Lagos: Longman) 1986, p.35.
13. *ibid.*
14. Figure obtained from Shell's Western Division, Warri. Since 1956, Shell never gave accurate figure of gas flared. The NNPC that ought to do so also relies on Shell for the figures. All has come to politicize the figures with the tendency of increasing the rate of gas consumed and reducing the figure flared. The Nigerian Gas Company is not any better.
15. Based on figures obtained from the Community files and Shell, a average was taken. This is because of the politics of the figures.
16. See No.UCH/48/73, Ughelli High Court delivered on November 12, 1974.
17. Based on the author's field trips to the area.
18. Based on the author's investigation in the area. The Youths of the areas cooperated. OMPADEC was hostile to the author.
19. See *The Guardian*, Lagos, 1994.

Chapter 8

Shell and Rural Land Crisis in Isoko Oil Areas

8.1 Introduction

To begin with, though then Abacha junta had maintained that Ken Saro-Wiwa and eight other Ogoni men were hanged on November 10, 1995 in Port Harcourt on murder charges, the main cause for their protest was to halt the pollution of their environment by Shell. Governments and non-governmental organisations from all regions of the world condemned their execution no doubt, but it was through that singular act that the global community, the parent Euro-American oil transnational corporations including even Shell International, got to know the depth of environmental crises caused in the Niger Delta by the operations of Shell Petroleum Development Company. More than anything else, the Ogoni episode unearthed the state's alliance with Shell in degrading the environment of the Niger Delta.

Given Oloibiri's experience where Shell operated for 30 years, pulled out after its wells became extinct, and left behind no social amenities but heavily polluted land and creeks, the Ogoni revolt was more less an attempt to check a repeat of such occurrence in Ogoniland. 200 kilometers west of Ogoni, the problem is not so much with Shell's pull out from Oloibiri as the trend of emergent structure of underdevelopment across the oil communities in the Niger Delta: creeks where no fish can be caught; lands that cannot support farming; and people whose cultural basis is destroyed etc.

Oloibiri is not only the oil community faced with this nature of ecological crisis in the Niger Delta region. Nor is Shell the only culprit as other oil companies like Mobil, Chevron, Elf, Texaco which came into the country's oil sector after 1960, are also guilty of environmental pollution as explained earlier. The only difference is that Shell came first into Nigeria's upstream petroleum sector, cornered most of the viable on-shore locations, and combined with its recklessness, still accounts for most cases of land pollution in the host oil areas of the Niger Delta.

Of all the oil communities where the experience of Oloibiri is being gradually reproduced in the West of the Niger Delta, Isoko area. Eighty kilometers west of Oloibiri, and 300 kilometers further west of the Ogoni community, the Isoko area was the second place where Shell found crude oil in 1958, about the same period it discovered oil in Ogoniland. The Isoko communities, host a large number of Shell's oil and gas fields on the West of the Niger Delta. The area also suffers from the worst ecological crises attendant to oil exploration and production.

Yet, the sudden currency of the 'green' issue in the global agenda for development has directed scholarly enquiries on the Niger Delta with emphasis on the Ogoni area as if it is worst hit. For one thing, the majority of the Ogonis are even as ignorant of the real issue at stake as most of the scholars who are only desirous of getting a paper published without a deeper idea of the root cause of the agony of the people living in the oil-bearing communities. Many officials of the Nigerian government have an understanding of the magnitude of destruction that Shell has done to all aspects of the environment and lives of the people in the oil host areas. This explains in part why no real thought is given to land, the most hit by all forms of pollution, of the State and Shell. In other words, it can be asked: Which land do the Ogonis still lay claims of ownership to, the polluted land or the crude oil found there? Should it be assumed that gaining full control of the land will mean both an end to environmental pollution and the resultant land crises there? Does anything suggest that should the Ogonis become part or full owners of the oil acreages, then the problem of land pollution will be over? Though the author's previous study of the indigenous oil companies have attempted to answer some of these questions,[1] they will still help shed light on the land crises. For one thing, if the Ogonis are to produce the crude oil found in their area, they will still be faced with land politics, particularly as familial and communal ownership of land and its resultant problems have never been given any thought. Even as the MOSOP continues with the struggle since the execution of nine of its members, no thought is still given to both local and national land questions and their attendant conflicts among and between the various communities that make up Ogoniland. It is for this reason that this Chapter analyses the experience of the Isoko oil areas to highlight this neglected but very crucial aspect of how Shell has underdeveloped its host oil communities in the Niger Delta.

Furthermore, the recent global concern for any environment inclusive of Nigeria's oil areas to sustain itself, has thrown up more political issues such as the right of the people to good environment. If the majority of Nigerians can no longer get food to eat, and unable to provide shelter for themselves, I wonder how those who live in the polluted oil areas can ever actualize their right to a good environment. Perhaps, it is only the well fed who should worry about environmental pollution. That is not to assume, however, that it is not an issue; it is by all standard. But sad enough, the emerging literature on the field has not really done much to examine the root cause of rural poverty including landlessness across the oil areas. Let us illustrate with the work by E. Osaghae,[2] about the most serious work done so far on the Ogoni revolt.

To E. Osaghae, the Ogoni crisis is largely a demonstration by minority people who are not just seeking for self-determination, but also, want to have control over their area including the natural resources notably crude oil; and right to fair representation in government. He argued that, the demand for the greater share of the oil wealth by the Ogonis, who bear the brunts of its production, stemmed from the increasing diversion of oil money by the central government to develop the non-oil areas most of which are in the northern part of Nigeria.[3]

Osaghae's position is well taken. However, one of the basic issues that remains unclear in his analysis of the Ogoni episode, is the difficulty of establishing real linkage between minority politics and the redress for environmental pollution. Important questions that will compel one to look beyond minority politics in understanding the pillage of the oil areas are: Should the minority oil areas including Ogoniland become republic, will that end the pollution of the environment? Is it to be assumed that the oil areas are backward, neglected and polluted because they are in minority? As prospects are high that crude oil may be produced soon in some parts of western and northern Nigeria, will the issue of minority still be relevant as the state continues to collude with the oil companies to pollute the environment? How will he (Osaghae) explain the degradation of the land of the host communities of the Ririwai minefield (tin) in Kano, one of the strongholds of the Hausa/Fulani ethnic group that has dominated Nigerian politics for many years? Is the problem not more with the character of the Nigerian state its mode of surplus extraction and conception of development than explaining the agony of the oil areas largely in the context of minority politics? Is the land question particularly its ownership, not at the root of the environmental problem? These are some of the fundamental questions which Osaghae glossed over in his work, but the case of study of the Isoko area will shed light on.

In all rural areas in Nigeria, land is central to the social and economic well-being of the people. The Isoko rural economy subsists on land, meaning the capacity of its people to reproduce themselves socio-economically, rests with the sustainability of land itself. Therefore, land, used here as an environmental resource and consisting of the earth surface, sub soil, air, water, and all other things like crude oil that is naturally attached to it and lying beneath it,[4] is at the root of the survival of the people. Taking its departure from previous works on the field, this Chapter, examines how the activities of Shell have caused and crises in its host Isoko oil-producing areas. It goes beyond the findings of Isichei,[5] Imevbore,[6] Oluwatimilehin,[7] Amajor[8] and Oteri[9] on the effects of gas flaring and oil spillage on the environment inclusive of land, to deeply examine land pollution and its resultant socio-economic crises, noting how they have affected the material well being of the Isokos. Finally, it analyses how the inhabitants of the Isoko oil areas have protested their exploitation and their emerging survival strategies, and responses from both Shell and the state.

How do the oil companies including Shell relate to land? Who actually owns all lands including those in the rural oil areas where the oil companies operate? It is to these and other related questions that the next discussion form.

8.2 The Land Question

The question of land ownership is about the most fundamental issue around which the politics of environmental pollution and its attendant socio-economic crises are centered in most African mineral-producing countries. Previous studies on the land question in Africa, notably those by D.R. Denman,[10] N.A. Ollenu,[11] J.A. Omotola[12] and S. Moyo[13] have shown that, land was no man's creation, but a gift from God. Not only that; land ownership was rooted on the basis of first occupier and its use over a long period which sometimes spanned twenty years or more without counter claim. That explained in part why in pre-colonial Africa, the structure of land ownership and tenurial system were familial and communal, though in some cases, traditional rulers and village heads held land in trust for their people.

With the imposition of European colonial rule, on Africa, the structure of land ownership was altered. Coming from political and socio-economic backgrounds where land was privately owned by individuals, groups and the state, the colonizing European powers took advantage of African traditional land tenure system to further their economic interests. European land laws were imposed on the colonies, laws which vested all land including the resources found therein especially minerals, on the colonial state. That also defined the content of land ownership in colonial Africa. The mining companies, therefore, only had access to land through the colonial state, and for which they paid rents and royalties to the state for the period the land was on lease. Since their acreages were on lease, the mining companies were not accountable to the people, but only to the state whose main interest was to get more rents from land and royalties for the exploitation of its resources land protection in terms of its sustainability was not on the agenda of the state. This was the brief account of the origin of how Africans were turned into squatters in their own native land by the colonial powers.

In Nigeria, the British colonial government adopted various methods to dispossess the indigenes of their lands. After the conquest of the Sokoto Caliphate by the British, they forcefully dispossessed Uthman Dan Fodio of all lands including those he had confiscated during the Jihad war. By colonial proclamations of 1900 and 1910, the power of emirs over all lands in a territory to be known as the Northern Protectorate of Nigeria, was revoked and vested in the state. In the South, where, the *obas*, *obis* and *ivies*, unlike their counterparts in the north, held land in trust for their people, individuals and families still had private right to land which they had farmed and lived on for about 20 years. It was this traditional egalitarian land ownership structure that largely informed the British gradual approach to land acquisition in the South especially through the use of treaties unlike the use of force in the North. Through the Treaty of Cession in 1861, King Docemo signed off the territory of Lagos to the British, which later became the Colony of Lagos in that same year. By treaties also, the Oil Rivers Protectorate and the Niger Coast Protectorate were created to consolidate British ownership and control of land and its resources in the Niger Delta where crude oil was discovered and still produced since 1956. The Southern Protectorate, a merger of the Oil Rivers and Niger Coast Protectorate in 1906, extended the frontiers of the state ownership of land and its resources as well.

By 1914, the Northern and Southern Protectorates were amalgamated to become what is known today, as Nigeria. In that same year, the Mineral Act previously formulated in 1912, was upheld to consolidate the state ownership of land and all minerals found therein in the country. By 1947 when Shell returned after the end of the World War II in 1945 to resume full scale exploration and exploitation of crude oil in Nigeria's Niger Delta, the Mineral Act of 1914 was still in force. As explained earlier, Shell was the only oil company permitted to operate in Nigeria, being a British colony. Since the Mineral Act was not meant to check land and environmental pollution of the host areas, mining companies operated with reckless abandon. Shell also took advantage of the situation and wreaked havoc on the environment when it began full operation in the Niger Delta. The situation has not changed particularly as its acreages, are regarded as minefields where oil spillages and dumping of untreated drilling wastes are normal operational field hazards.

After Nigeria's political independence in 1960, the Mineral Act of 1914 was inherited with major changes that replaced clauses like Crown Government with the Nigerian government. Land and minerals found therein, were all vested in the Nigerian state. Like its predecessor, the state has no interest to deal with land pollution. Rather, it took a considerable portion of then Mineral Act of 1914 into the Petroleum Act of 1969 to consolidate its accumulative base in the burgeoning oil industry. All that encouraged Shell and other oil TNCs to pollute the environment/land of the host areas. With its greater rate of land degradation by the oil companies, the state became ever sensitive to protest over land pollution in the oil areas. Accordingly, it enacted the Land Use Decree of 1978 to reaffirm its sole ownership of lands including those oil is explored, produced, transported and stored. The decree also protected the interests of the operating oil companies by shielding them from land litigations.

As land pollution deepens with its worsening attendant socio-economic crises in the host oil areas, the people who live there have been challenging the state's sole ownership of all lands, rural and urban. Communities have been at war with each other. Bloody conflicts over land are however, not limited to the oil areas. The protracted war between the Tivs and Jukuns over land has cast doubt on the supremacy of the Land Use Decree of 1978. J.A. Omotola, one of the strongest critics of the Land Use Act, has described it as alien, ill-conceived and deepening the poverty of the rural people, particularly those in the oil areas of the Niger Delta, who have been forcefully evicted from their own land by the state. To him, the raging ethnic conflicts over land in rural Nigeria is an indication that, the private right of the people to their land is still an issue. On its part, the state is not finding it easy to implement. J.Obaseki, a one-time Judge of the Supreme Court of Nigeria, summed up the status of the Land Use Act in the country as follows:

> Land may now be easy for the government to acquire, but for the average Nigerian or common man, it is a lost hope or objective. The claim for title to land has not diminished. All that has happened in the Southern states is that the claims have shifted from title to claims for entitlement to rights of occupancy. The Courts are no less busy than before in trying to sort out the competing claims. There is dire need for reforms in the Law of Property in Nigeria. A total overhaul and re-examination is necessary.[14]

However, as the politics of oil in the country throws up a wide range of conflicts over land, nothing suggests from all agitations in the oil areas that the people really want to stem the pollution of land including the larger environment. As the Isoko experience will soon show, leaders of agitations capitalize on protest to further their own political and economic interests leaving land crises to deepen.

As the battle over land rages on between the state, the oil companies, and the people; and among and between the communities in the host oil areas, the operating oil companies not least Shell, exploits the confusion to its own advantage. How is Shell implicated in the seeming land conflict?

8.2.1 Nature of State-Shell Alliance and the Implications for Land

As stated, the upstream sector of the Nigerian petroleum industry was until 1960, monopolized by Shell. And during that period, both Shell and the colonial state had one corporate economic interest in Nigeria's Niger Delta: to produce crude oil profitably without caring for land. In essence, while the state provided the enabling legislative environment, for capitalist accumulation in the industry, Shell produced the crude oil. Worst still, neither did the policies of the state, nor the nature of the operations of Shell create the necessary domestic linkages with the economy, especially the development of the vital productive forces like skilled labour on which the country's oil sector would be anchored after 1960. All was to enable Shell maintain firm grip of the country's oil industry in the colonial period. This continued into the post-colonial era. As to be expected, though the Nigerian state owns the oil and gas sector, it lacks the technical capacity to operate the sector, making its control of the foreign oil companies like Shell. Incapacitated by a weak technological base, and lack of an energy policy, the state's intervention into the petroleum industry has largely turned out to be for rent collection. In part, this explains its inability to tackle environmental/land degradation of the oil areas by the oil companies generally, and those of the Isokos by

Shell in particular. All this informs the nature of the state-Shell alliance and the politics of land pollution.

Shell has all its acreages and lands in which they are located leased to it by the state initially for 99 years until in the 1980s when it was reduced to 33 years as explained earlier. The vital point to note is not the long period of lease; rather, it is the context in which Shell operates on such land, a minefield. This explains the company's rapacious exploitation of the land for the lease period. It became worse since 1973 when the NNPC/Shell joint agreement came into force. For, not only does the state like Shell, conceives of the host oil areas as a minefield, the company still handles all operations for the joint venture, leaving the state to collect. Under this kind of arrangement, the state is being left with one central to role to play: to guarantee regime of accumulation. This is because any threat to the operations of Shell will, in turn, bring about drastic decline in the revenue accruing to itself and the company. This, as shall be made clear soon, is to further empower Shell to degrade land. Under the joint venture agreement too, and to which Shell is the technical operator, back-passing between the state and the company as to who actually polluted land and the longer environment has been the rule of the game[15]. All has fueled agitation by the people who live in these oil areas. From all indications, the joint venture between NNPC and Shell is a hoax, and its hollowness is evident in the worsening land crises across Shell's host oil areas with the Isoko communities worst hit. But before looking into the consequences of Shell's activities on their land. Let us understand who the Isokos are.

8.3 Brief History of the Isokos and their Land

Historically, the Isokos are descendants of the old Benin Kingdom. Oral history dates the arrival of the Isoko people in their present location in Delta State to 1870. With Isoko as their language and numbering 276,395 people spread over an area of 1,048 kilometers,[16] the Isokos constitute one of the minority ethnic groups in the Niger Delta region of Nigeria. The Isoko Council Area, (North and South) has 17 clans namely, Emevor, Ellu, Irri, Ofagbe, Ozoro, Uzere, Emede, Eroha, Iyede and Okpe. Others are Owhe, Igbide, Olomoro, Aviara, Oleh, Enwe and Uweh. The Isoko area is located within the Equatorial zone, precisely longitude 6° 25' East of the Greenwich Meridian, and latitude 5° 51' North of the Equator which is subsumed in the location of the oil-rich Niger Delta as explained earlier. It shares boundaries with the Orogun and Kwale people who are to the north; Ughelli and Agbarha communities to its north-west; the Ijaws in its south-west; and the

Ase River, a tributary of the River Niger, to its south. Farming and fishing are the predominant occupations of the people.

The Isokos, like most people who live in the rural aeas of Nigeria, attach significant sociological importance to rural land since it traces their historical and cultural roots. Among the Isokos, too, land ownership and rights of title to land are of multiple nature. As republicans, heads of communities only hold communal lands in trust for their people. Traditionally, lands can be classified into three types in the area namely, communal, family and individual. In every community, it is the *Ovie*, the *Odion* who is usually the eldest man, and chiefs, who preside over all land matters particularly land sharing among the people. It is from communal land that each *quarter* or part of the village, is allocated land, from which an *odion*, in turn, shares among the major *houses* or a group of families that makes up that quarter. Finally, is the family land, which the eldest man in the family shares among the only married male adults. It is this plot of land that belongs to the individual. However, the land sharing exercise is a continuous one. As more male adults get married, and bear children, the more the need arises for them to get their own land. Also, as heads of families die, so are their lands reallocated to the eldest male children in order to avoid land disputes given the spread of polygamy in the area.[17] Male children borne out of wedlock are traditionally not allocated land following the death of their father. They are agitating for the abandonment of this tradition so that they can get a share of the land. In fact, this is a major sources of family land conflict in the area which is escalating as a result of land pollution.

The right of the people to land is by inheritance, passing only from father to son or the eldest male children in case of polygamous families. Land sale is, therefore, traditionally forbidden. In cases where land is pledged, its ownership by tradition, is not transferred. Rather, it is the right of use to such land that is on lease, the ownership reverses to the debtor after liquidating his debt. Land holding size is usually small, about an average of 800 square meters for a family of four consisting of the husband, wife and two children. Though the holding is small, soil fertility is high and each family has five other farm plots of about the size to till, meaning there was no land hunger until Shell began operation there. With the exploration and exploitation of crude oil, land ownership pattern was upturned, and its traditional use and management system destroyed.[18] Instances abound in the Isoko area where the three types of land have all fallen within Shell's acreages, leaving the people with no other option than to farm the polluted minefield side by side with the company's mining activities. Since the company's operations cover 15 out of the 17 clans in the area, the resultant land pollution is widespread. More details on the consequences of Shell's operations for land are taken up in the course of the discussion. A brief look into the impact of British colonialism on rural land in Nigeria will provide a clearer idea of how the Isoko area was particularly affected.

8.3.1 British Colonialism and Rural Land in Nigeria: A Brief Analysis of How the Isoko Area was Opened up for Shell's Exploit

British colonialism did not have common land policies in all its former colonies in Africa; nor was its interest on land the same in these colonies. While it did not have vested interest in commercial farming in Nigeria and therefore, did not grab all fertile lands contrary to what it did in Zimbabwe, among other countries in Southern Africa; and in Kenya where it built to settler colony and prime lands were cornered for commercial agriculture. Whatever the different uses into which land was put in the colonial period, the state was the sole owner of all lands, rural or urban. Let us dwell more on this with the impact of colonial land policy on land use pattern.

As an agricultural and mineral producing colony, colonial policy on land in Nigeria was not the same in the sense that while land in the mineral producing areas was directly taken up by the state thereby affecting both its tenurial and use systems, the rural peasant farmers were still allowed to till such land. In terms of agriculture, their own traditional farming system was encouraged particularly as the state was not involved in commercial farming and therefore, relied on them for produce. In its attempt to get more agricultural produce for the metropolitan factories, the colonial government enacted the Native Rights Act of 1956 which, further, consolidated land on the colonial state, but only made the traditional rulers its custodians.[19]

In the mineral sector, the situation was different. As explained earlier, the colonial government was only interested in the collection of rents from the foreign mining companies. Tin was mined open cast, and coal by underground. Both methods of mining had attendant negative consequences for the traditional land use systems of the host areas. Either method took a toll on farmland, forcing farmers to extend the frontiers of land hunger in the host areas of Jos, Kano for tin; and Enugu for coal as they search for arable land. As it shall soon be made clear, the production of crude oil in the Niger Delta aggravated land problem as the indigenous methods of tilling land and its management were uprooted.

During European colonial rule in Africa, the continent was invaded and its people were treated as if they had no culture. In Nigeria, the British navy led the incursion, destroying all the belief systems of the people. The British mission was not just to conquer the people, but to dispossess them of their land and have their culture destroyed. Christianity played more of the latter role than preaching the gospel. As Eme Ekekwe rightly noted, both the gun and the bible served the same purposes: to facilitate the exploitation of the colony.[20] I share his position moreso as it has relevance for land. The origin of land pollution by the oil companies in Nigeria can be traced in part to the destruction of the sociological significance of land following the advent of European christianity. Christian doctrines for instance, encouraged the people to eat totemic animals. With more converts, revered animals were freely killed and eaten. Most of the rites hitherto observed for them were abandoned. Churches were built on 'bad bushes', or traditional cemeteries where all socio-economic activities were prohibited. Virtually all traditional myths on land were destroyed.[21]

In the Isoko oil areas, like most rural communities in Nigeria, however, not everyone was a convert. Some of the people still traditionally observe the rites for land. But it is like an exercise in futility as the foundation has long been damaged even before Shell began operation there. Shell has no respect for the belief system of the people. It is with such utter disregard for the culture of the people that it explores and produces crude oil in the area. It does not matter to Shell whether oil was found in a 'bad bush'; for graves of the dead are mobilized in its search for oil.

Though British colonialism did lay the foundation for exploitation of the Isoko area by Shell; on its part, the company has taken undue advantage of its collaboration with state to cause more serious damage to the environment and land. It is to the detailed analysis of Shell's operation and their consequences for land that the next section now turns.

8.4 Consequences of Shell's Operations

Before the discussion proper, it is important to have a brief idea of the history of Shell in the Isoko oil area.

8.4.1 Shell in the Isoko Area: A Brief History[22]

Shell pioneered and still dominates the exploration and production of crude oil not only in Isoko areas, but across the Niger Delta. In 1957, the company had its first seismic shot at Uzere, began drilling there the following year, and finally, struck commercial wells in 1959, just about the same time that it also found crude oil in Ogoniland, and the Ughelli oil areas. Within the period, too, Shell discovered oil in other Isoko communities, notably, Igbide, Emede, Olomoro, Ogini, Owhe, Oroni, and Ilue-Ologbo. By 1960, the Isokoland had become and still remains one of Shell's largest on-shore locations in the Niger Delta as evidenced in the number of oil wells it operates in the area. At the time of writing, Shell has a total of 102 wells located as follows: Uzere (34), Olomoro (33), Owhe (II), Ogini (17) and Oroni (7). Shell's activities are more concentrated in the Isoko are compared to Ogoniland where it has 58 wells, in all; and even less in the Oloibiri field with 12 wells before they became dry. Since 1959, the company has produced a conservative total estimate of 200 million barrels of crude oil, and close to a trillion cubic feet of associated gas flared from its various oil/gas fields located in the area. It maintains six flow stations. One each located at Uzere East, Uzere West, Olomoro, Owhe, Ogini and Oroni. Through a network of pipelines, the company links all its oil/gas fields to the flow stations where *gangue* materials like water associated with crude oil *in situ*, are first chemically separated before it is pumped to the company's only Central Quality Control and Pumping Station at Ughelli, which is outside of the area, but only about 30 kilometers away. This station serves all the company's other oil fields throughout the west of the Niger Delta. It is at the Quality Control Centre too, that the crude oil undergoes further processing in order to ensure that other chemical sediments are removed before it is pumped to the Forcados terminal from where it is finally exported to Europe, USA and the rest of the world.

As explained earlier, Shell's access roads in any oil community including that of Isoko, is primarily to facilitate its operation. Though the people might social and they are not originally planned to help bring about the economic development of the area previously discussed in Chapter Seven. For, as soon as Shell's oil well becomes extinct, it abandons the access roads as in the case of Oloibiri. Nor does it extend the road to any community that has no oil even if it is less than a kilometer as in the case of Imodje-Orogun under the Ughelli Council Area. Not even the state is interested in taking up such tasks. Now back to the discussion proper beginning, but first with Shell's mode of operations. What is Shell's relationship with the OSCs?

8.5.1 An Analysis of Shell-OSCs Relationship and Implications for Land

As stated earlier, Shell is the technical operator of the NNPC/Shell joint venture. But that does not mean all oil exploitation is done by Shell. Rather, it is the OSCs that do the actual exploration, production, transportation and storage of the crude oil for Shell. To that extent, it can be argued that Shell is no less than a contractor. In fact, that has been the trend in the global oil and gas industry when the Anti-Trust Law of 1911 came into force in the USA. If Shell is a contractor, the question is: Should the OSCs not be held responsible for the degradation of the environment? Not really. As explained in Chapter, the OSCs only operate for Shell, meaning they are not liable for the ecological crises arising from oil production in the Niger Delta. Instead Shell does. That encourages the OSCs to operate with reckless abandon. This is all the more so as the state, sole owner of all lands, prepared to halt land pollution.

On their parts, the oil servicing companies are only contractors to Shell and nothing more. They do not owe the host communities any obligation much more protecting their land. They do not relate to the state, but only to Shell. As contractor and servicing companies, their aim is to find and win crude oil for Shell. And as a consequence, they wilfully pollute land, believing that the state and Shell will later handle its effects. This is the context in which the OSCs operate on behalf of Shell and other oil companies in the Niger Delta. As neither the state nor Shell has ever taken environmental protection serious, the servicing companies will continue to pollute the environment and continue to go unpunished for a long time to come[23]. A tread the youths are fighting to halt by first preventing the OSCs from prospecting in the area. This will become clearer as the actual operation of Shell are examined.

8.5.2 Exploration[24]

As the first stage of finding crude oil, the assault on land begins with the pathways constructed by the seismic crew of the OSCs. But the actual process of land degradation begins with the use of chemicals like pipelax, bentonite, EP-20 used during seismic and pilot drilling; chemicals that are not only harmful to plants and animals, but whose untreated dangerous wastes, all dumped on farmlands and into swamps are further degrading the ecosystem of the host area. In one of the seismic operations at Isoko oil area by Seismograph Services Limited, the another observed how, tramps (cables) were laid over an average of 10 hectares of farmlands, and across creeks regardless of human activity. And for two months that the operation lasted there, all economic activities on the affected land were first disrupted before coming to a halt. Furthermore, after several seismic shots in the delineated area, preliminary drilling showed that the underlying rocks were oil-bearing. Even if it were a dry hole, meaning no crude oil was found, not less than 15 hectares of land were found to be polluted. Other traditional rites attached to land such as no-shooting of guns or detonation of dynamites further violated aspects of the culture of the people that has to do with the war deity, burial and god of land. Because the area was found to have huge deposit of crude oil, more explosives were detonated in order to commence actual drilling during which an additional 15 hectares of land were lost to the company. In all, 30 hectares of farmland were lost to Shell in a single seismic operation.[26] How many families were displaced in this operation?

In the past, the size of farmland per family was an average of about 800 square meters, that is, 400 meters long by 200 meters wide. That is before the advent of Shell. In computing this area of land traditionally, a man of an average height of about 4.5 feet is taken as the unit of measurement. His hands are stretched out sideways, and the length which approximates his height, becomes the standard unit of measurement. This, he does with each point marked off for about 89 times to give the length, and 45 times for the width. In spite of the influence of western education, this aspect of the culture of the Isoko people particularly as it applies to the sharing of land still persists. Taking the 800 square meters or 0.8 hectares land holding, per family, it means that for every 30 hectares of land lost to Shell, about 38 families were affected leaving not less than 150 people displaced from their farmlands.[26] But this figure is not sacrosanct; nor does it give an idea of the depth of land pollution and the attendant land crisis in the area. For the real crisis resides with the continuous loss of farmlands due to intensive explorations in the area, which, in turn, has left more families without an arable land. In fact, as more oil wells age, Shell renews exploration of hitherto abandoned fairly minable deposits in the area. Suffice it to note that, as exploration is a continuous process, so also, are its polluting effects on land unending.

It is difficult, therefore, to give absolute figures of all lands, be they farmlands or creeks/swamps, lost at this stage since they are temporary and only for limited period. For instance, using the 0.8 hectare landholding size for a family, about 420 hectares of farmland were lost to Shell between 1957 and 1958 at Uzere with 525 families affected. About 2100 people would have been displaced. But the Uzere community has suffered more loss of land from exploration since then. Even oil has been prospected in most of the farmlands for as less than five times, making it all the more polluted and therefore, of no use documenting the situations in Uzere and other communities in the past 40 years of Shell's exploit in the area. However, as of 1997, not less than 20,000 families in each clan had lost their farmlands though exploration since the advent of Shell.[27]

No doubt, oil exploration took and still takes toll on land in the Isoko area. But under production, more lands are rather permanently used up thereby heightening landlessness and its attendant social and economic crises. Before looking into the implications of oil production for land, it is important to note that in some cases, the same land being previously explored is subjected to another round of pollution when production begins there. All is to deepen the incapacity of land to reproduce itself.

8.5.3 Production

In producing crude oil, Shell, like other oil companies, uses definite hectares of land. This is in addition to other parcels of lands polluted in its previous operations. Crops are also compulsorily destroyed in the process of oil production. Let us begin with the specific hectares of land lost during oil production. Table 8.1 gives an insight into the hectares of land used up in the various aspects of oil production.

Table 8.1: Nature of Oil Activity and Land Lost

Production Processes/Activity	Land Lost
Oil Well	6.3 hectares per well
Flow Station	6.4 hectares per station
Camp	1.2 hectares per camp
Helicopter	1.4 hectares per field

Source: Shell, Western Division, Warri; Author's Field Works in 1993 and 1995.

In addition to hectares of land lost to the process of producing crude oil as contained in Table 8.1, land is also lost to Shell through its right-of-way on which its oil/gas pipelines are laid. Right-of-way usually has an average width of 500 meters and all lands within this gauge are automatically lost to the company. Its length depends on the distance between the various oil/gas field, quality control centers and terminals. Details on this are treated later. Oil wells and flow stations have more direct impact on land as shown in Table 8.2.

Table 8.2: Hectares of Land Lost to Shell's Oil Wells and Flow Station

Location	Nature of Operation and Land Lost		Total
	Oil Wells	Flow Stations	
Uzere East and West (2)	34 x 6.3 *214.2 hectares*	2 x 6.4 ha. *12.8 hectares*	227 hectares
Olomoro			2214 hectares
Owhe	33 x 6.3 *207.9 hectares*	1 x 3.4 ha. *207.9 hectares*	113.5 hectares
Ogini	11 x 6.3 *207.9 hectares*	1 x 6.4 ha. *+6.4 hectares*	60.5 hectares
	17 x 6.3 107.9 hectares	1 x 6.4 ha. *1 x 6 hectares*	
	Total		**680.7 ha.**

Source: Author's Field Work, to the Area in 1995; Estate Department, Shell, Western Division Warri, 1995.

As shown in Table 8.2 above, 681 hectares of land not previously polluted by exploration, were permanently lost to Shell. However, this figure was only a rough estimate of the actual hectares of land lost during oil production. For, within the site of a typical oil well, an additional 2 hectares of land were allowed for space. Yet, such parcels of land were not added to the total land being used by Shell. Which means an additional 200 hectares of land compulsorily lost in the process of crude oil production were not accounted for in the company's records.

For flow stations, approximately 3 hectares of land were consumed by the 'No Trespass' order of Shell/NNPC, meaning an additional 18 hectares of land were lost. As of the end of 1996, about 900 hectares of land were lost to the company's oil wells and flow stations. Using the 0.8 hectare farmland per family of four as unit of measurement, about 4,500 people were directly affected.[28]

On gas flaring, Shell has deliberately refused to document hectares of land on which its flaring sites are built. It is all part of the company's plot to deny the people knowledge of the exact hectares of land being wasted through flaring of gas since it can worsen their anti-Shell feelings in the host areas. But that is a mistaken position by Shell. For, the youths have, in conjunction with independent researchers and some of its officials in the Estate Department, unearthed the guided secrets. Taking the flaring point as the 'epicentre', one square kilometer of land has been found to be affected in one gas flaring site. In larger gas fields such as the one at Eruomukowharie in the Ughelli oil areas, with four flaring points, about 3 km^2 of land is used up. On average, not less than 100 hectares of land are found to have polluted per site. This is in addition to the heat from the flame which causes general warming and acid rain from the dangerous chemical compounds like carbon monoxide emitted into the air. An additional 200 hectares of the surrounding land and creeks have also been found to be polluted. In all, about 300 hectares of farmlands were polluted in the host oil-bearing areas of Isoko. Again, using the 0.8 hectare of farmland per a family of four, that translated into an average of 375 families directly affected with 1500 people displaced from their major economic activity, farming. From the company's six flare sites in the area, about 1,800 hectares of land were found to be polluted, leaving 2,250 families without farmlands and over 9000 people thrown out from other land-related economic activities.[29] In addition to the disruption of economic activities of the people, gas flaring has also posed health danger to the people. As of 1997, bronchial asthma and miscarriage of pregnancies among women were some of the reported cases becoming prevalent in the area after long period of exposure to the heat from the flare. The trend is likely to continue into the next millennium. One other crucial point to note on the implication of gas flaring for land is that, any land that has been used as a flare site is almost permanently unfit for farming due to its desiccating effect on the soil that will last for about 30 or more years. Creeks, too, suffer the same problem. This explains in part why there are patches of 'derived savannahs' littering most of the company's previous and present locations in Isokoland. Even if there is a stoppage of gas flaring now, which is unlikely, about 30 years will be needed for such polluted land to renew itself with the application of fertilizers. But land in Isoko area, and any other oil-becoming communities in Isoko area can hardly be given such time as exploration intensifies.

Shell's camp sites in the Isoko area, like in other oil areas, are essentially used for the storage of heavy earth-moving equipment, drilling tools and make-shift accommodation. Precisely, that is what the 1.2 hectares per camp as shown in Table 8.1 represents. But this figure is quite misleading as it does not include other make shift accommodations erected by the numerous casual workers of Shell who use up considerable hectares of land. Nor does it give an idea of the parcel of surrounding land that is wilfully assaulted by the oil workers as well. Putting all this together, 5 additional hectares of land were found to be used up by Shell, bringing the total to 6.5 hectares of farmland.

Furthermore, it is instructive to note that, camp sites are temporary. So, too, is the use into which such land is put. Forty days, about the longest period the oil workers stay in such temporary sites, might be considered short to cause any land crises. But the magnitude of the assault on land in that period was found to be usually grave that such land would need no less than a fallow period of 5 years to renew itself. But that cannot be guaranteed since other aspects of oil production are taking place at the same time on such marginal lands. Worse still, the contractor service companies to Shell have never reclaimed swampy land for camp site, rather, they use existing dry farmlands since they are easier to mobile. To that extent, the land used for camp sites has deepened land hunger and its attendant socio-economic crises in the Isoko area just as exploration, drilling, gas flaring, and oil spillage are degrading the environment in greater intensity. By 1988, each of the 15 clans had lost about 120 hectares of land most of which were farmlands through camp sites, bringing it to a total of 1800 hectares throughout the Isoko oil communities. 9000 people were displaced.[30]

In order to construct its right-of-way on land, Shell obtains legal permission from the state and for which it pays fees. Until 1988 when the period of lease was reduced to 33 years following the adoption of the World Bank/IMF-induced adjustment by the state, it was 99 years. Of the company's 2000 kilometers network of oil pipeline across the Niger Delta, 200 kilometers are in Isoko land. About 1.5 hectares of rubber plantations were affected in 20 out of the 50 villages in the area that the pipes passed through. Uzere, Oroni, Ellu, Ilue-Ologbo were among those communities particularly badly hit. In all, 70 hectares of farmland and 30 hectares of rubber plantations were directly affected. No fewer than 125 families were displaced.[31] Because the people have no other arable land to farm, they willingly disobey the orders of the NNPC/Shell on such land, and continue to till such marginal land where oil exploration and production are going on at the risk of being killed by the explosion of the high pressured oil/gas pipes. Details on the survival strategies adopted by the people are taken up in the course of the discussion.

At no other stage in the finding, mining and transporting crude oil has the recklessness of Shell been so more profound than in the frequency of incidents of oil spillage. Not only are the company's oil pipelines still laid bare on the ground in its host oil areas in the past 40 years, they are without any routine maintenance. That partly explains why ruptured oil pipes still account for 70 percent of cases of oil spillage throughout the Niger Delta including those in the Isoko area. Shell is not alone in neglecting its flowlines, Mobil, Texaco, Chevron, Agip, Elf are also guilty of the same recklessness. Let us illustrate with the specific case of the Isoko area noting its impact on land.

8.5.4 Oil Spillage

Studies have already shown that crude oil, when spilled on the environment, pollutes it including land. It also destroys the nitrogen cycle of the soil, halts photosynthesis in plants, and kills plankton, fishes and other aquatic organisms.[32]

Being the first and largest oil producing company in Nigeria, Shell has come a long way. The company feels too powerful for the Nigerian government to control due in part to the latter's technological weakness, mode of surplus extraction from the industry, many other problems. All this explains why it operates in defiance of 'the ethics of good oil fields practices' in the Niger Delta especially in caring less about the mechanical and physical fitness of its flowlines. Of all the identified causes of oil spillage in the oil areas, ruptured oil pipelines accounts for 80 percent of the incidents with the remaining 20 percent caused by engineering error, poor maintenance, human error, sabotage, natural, like sand cut, and unknown. Not only that. Of the 10 operating oil TNCs in the country's oil and gas sector, Shell still has the highest cases of oil spillage because the company has both the largest concessions and about the most reckless. Reason for this has been explained earlier. And from about 600 incidents of oil spillage in the Niger Delta reported in 1985, and during which over 5 million barrels of crude oil were spilled and all lost to the environment, Shell accounted for 70 percent in both spills and volume of oil lost. By 1990 when additional 700 cases of spillage were reported, Shell also accounted for 75 percent. All this was indication to first and most disastrous oil spillage ever in Bomu, Ogoniland, in 1970 in which over 3 million barrels of crude oil were estimated to have been lost to the environment.[33] That is on a general note.

The first case of oil spillage in the west of the Niger Delta was recorded in 1973. Again, it was caused by Shell in its operational locations 13 and 18, at Igolu in the Isoko oil area. About 40,000 barrels of oil were estimated to have been spilled and all lost to the environment. Neither Shell nor the state cleaned up the oil slakes. Among the immediate observed impact of the spill, was the severe pollution of the underground water. Eight hundred fish ponds and 200 lakes where most of their fishing took place at the time of study, were particularly badly polluted as dead fishes among other aquatic organisms, were found floating on the water. 1000 hectares of family-owned rubber plantations were polluted which resulted in the eventual death of this economic tree. It also displaced the people of the affected area from their economic activity. All was to force about 4,000 people out of the total population of 5000 of the area to flee to the neighbouring Araya and Oleh communities where the people themselves, were already grappling with land hunger and rural poverty.[34] Shell did not conduct any post-spillage research in the area till date, nor did the state. In fact, the general attitude of Shell and other operating oil companies like Mobil, Texaco to oil spillage is that, they are not ready to own up to their recklessness, much more prepared to clean up and restore normalcy to the economic activities of the affected area. For instance, the Finiwa blowout in Bayelsa State, in 1981, in which Texaco and Chevron spilled an estimated 2 billion barrels of crude oil, was also repeated at Igolu. And in 1985 when Shell's oil and rig off the Bonny area sank and killed all the crew members, nothing was done to recover both the crew and equipment that were drowned; nor efforts made to assess the damage done to the marine life of the area.[35]

In 1980, Shell recorded yet another serious spillage at Uzere in which 100,000 barrels of crude oil were spilled with no single barrel recovered. As to be expected, neither Shell nor the state carried out any post spillage research in order to determine the extent of damage done to the environment and also, to restore the economic activities of the people disrupted by the spillage. While the Uzere people were yet to recover from the impact of previous spills, it was hit again by another major ruptured flowline in July 1983. It was the worst spillage ever recorded by the company in the area with about 200,000 barrels of crude oil reportedly spilled and all sank into the soil, polluting crops, traditional fish ponds, lakes and creeks. As a predominantly fishing community, 2,000 fish ponds were affected. Pollution was so severe that, by 1995, that is, 12 years after, about 6,000 family farmlands of the 0.8 hectare size per family, estimated at 4,8000 hectares were still found unfit for growing crops. An additional 2,500 hectares of rubber plantations directly affected by the spill, had their trees gradually dried up, giving way to an emerging 'derived savannah' in the area by 1995.[36] Because of the magnitude of environmental/land pollution caused by this singular spill in the area, Eniola Adeniji *et al* conducted an independent research to evaluate the impact of the spillage on both the surface and sub-surface water moreso as the Uzere community had no pipe borne water at the time of spillage. Table 8.3 summaries their findings.

Table 8.3: Conditions of Water Resources Before and After Oil Spillage on Uzere in 1983

Water Resources	Drinkable Water		% of Oil Pollution
	Before (1)	After (2)	
Water	7	7	100.00
Stream	19	10	80.30
Lake	12	10	83.30
Spring	14	14	100.00
Well (hand dug)	2	13	44.00

Source: E. Adeniji *at al*, in *Petroleum Industry and...* (1985).

From the author's most recent survey of the Uzere oil community particularly its swamps and lakes, the impact of oil pollution the wetland was still grave. Table 8.4 sums up the observed effects of the cummulative impact of oil spillage on the ecology and its resultant effects on fishing.

Table 8.4: Observed Effects of Crude Oil Spill/Pollution on the Uzero Wetland, 1984-1945

Company	Causes of Spillage	Other Mode of Pollution	Impact on Swamps/ Lakes	Remark
Shell	Ruptured flowlines	Dumping of untreated and toxic drilling wastes into swamps; Cutting well formation, and gas flaring	Trees mortality, fissuring, defoliations, decimation of mollusc, plankton	Catfish almost, extinct; fishing virtu- ally grinding to a halt.

Source: Author's field works, 1989-1995

As the community hosting the largest concentration of Shell's on-shore oil wells est of the Niger Delta, Uzere were been the hardest hit by oil pollution. And the situation will remain so for a long time to come. Other oil areas like Owhe, Ilue-Ologbo, Emede, Igbide, Olomoro and Ogini have also suffered from frequent cases of oil spillages. All has caused prevalent land crises in the area. These is nothing to suggest that a stop would be put to this trend as most cases of oil spillages in the 1990s actually occurred on already heavily polluted lands. A blow-by-blow account of oil spillages and their social and economic effects on the economy of Isoko area will be repetitive. Rather, it is the general trend of landlessness and the attendant worsening rural well-being of the people who live there that are now examined.

8.6 Consequences of Land Crises for the Social and Economic Well-being of the Isoko People[37]

To start with, Shell had through its activities since 1958, polluted virtually all land in all its host areas in the Isoko community at the time of this study. As to be expected, land crises were pervasive, though the magnitude varied from one community to another. While some of the badly hit oil areas like Igolu was on the brink of extinction, skeletal farming still took place on the marginal land. The land hunger in the area is traced to the state. As explained earlier, since all lands where oil is explored and produced, are vested in the state, the people who live in the host oil areas are only squatters in their own ancestral land. More importantly, these lands are regarded as the company's minefields where pollution is allowed. Herein lies the politics of the landlessness and the attendant rural poverty in the Isoko oil-bearing, area, among others.

Rural land in Nigeria including the Isoko oil communities, provides a safety net for the people. As a predominantly farming community, the people engage in subsistent agriculture, but not without a hand in other vocations like traditional medicine. Because of the widespread pollution of the environment, herbal plants like common club moss was almost extinct in the area at the time this study was conducted. As a result, native doctors in the area, like their counterpart in Ughelli oil could no longer rely on native medical practice to supplement their living and provide traditional health care for the people. Petty trading was almost abandoned. This is because poor harvest has decimated the quantity and quality of farm produce for sale, which used to be one of the main sources of capital for petty business. One of the results of all this, is increasing poverty, misery, hardship, vulnerability and hunger in the land. Some specific illustrations will be helpful at this point.

By the provision of the Land Use Decree of 1978, sole ownership of all lands where oil is explored and produced is vested in the state. As a consequence, unofficial new boundaries were constantly drawn by Shell, the state, and the inhabitants of the area to reflect the expansion of the company's minefields on the one hand, which in turn, has led to the increasing reduction of family land holding on the size. For instance, there has been a sharp decimation in the size of family land holding, especially since the 1980s when all the attendant effects of environmental pollution have fully manifested. From an average of 0.8 hectare of farmland held per family throughout the 1960s and 1970s, when the impact of oil pollution was yet to fully manifest, it declined to between 0.05 and 0.2 hectares since the 1980s with signs for further decreases across the community following commulative degradation of the environment. While most families at Uzere and Olomoro were found not to have even the already polluted farmlands to till, in Owhe, Ilue-Ologbo and Iyede, where marginal lands were found, the size of the holding was estimated at 0.2 hectares. The extinction of anthills was noticeably widespread in all communities in the Isoko area, an indication of acute land infertility. Cassava, the most staple crops in the area, suffered and still suffers from stunted growth and therefore, very poor yields. Other crops like pepper, cocoyam, plantain were equally adversely affected by pollution. Yellow yam, the most commonly produced specie of yam in the area, could no longer be grown. All is an evidence of protracted land pollution.

In no other areas have the cummulative consequences of landlessness and poor yield become so profound than in the living standard of the people. The acute hunger in the land has manifested in the drastic price increases of basic foodstuffs like garri, yam, plantain, pepper, since they could hardly be grown in the area. From the prices of foodstuffs sampled at different community markets in Uzere, Ozoro, Owhe, Ilue-Ologbo, Iyede, Igbide between 1990 and 1997, a 25 kilogram sack of both white and yellow garri, was sold at any average of N700, an increase of N400 over the 1989 price of N300. By 1993, it rose to N1000, and further shot up to N3,000 in 1997, and has continued to rise steadily since then. With the erosion of the means of income of many families, only very few of them could afford to buy garri at those prices. In fact, abject poverty still pervades other rural oil areas like Umuechem in the East of the Niger Delta where Shell has been operating since 1958, about the same time it came to the Isoko area.

With respect to fishing, though Uzere was particularly badly hit by poor catch, other coastal fishing communities like Ogini, Igbide were not spared as well. The emergent trend of fishing in these communities, was the gradual reproduction of the Oloibiri field's communities experience. As a result, the scarcity of fish, main source of protein in Isoko area, was persistent. Hence, its prices became very high. In 1995, an average size of matured smoked eel weighing about 500 gram was sold at N5,000 as opposed to N3,000 in 1994. Five average sized mud fish sold at

N700 in 1994, rose to N2,000 by 1995. At this price, only very few families could buy; and there were no indications that a majority of them would be able to buy as their main sources of income particularly farming and tapping rubber were almost impaired. Prices of fishes would not drop because they were not caught in the area. Although most families bought frozen fishes marketed by the Ibru Sea Food Organisations among the few fish trading companies operating in the country, the price of fish was still high. A one kg cartoon of pilchard was sold at N6,000 in 1996, a price far beyond the reach of the rural poor in Nigeria inclusive of those in Isokoland.

It can be discerned from the foregoing analysis that the socio-economic well-being of the people who live in the Isoko area is on the decline. Given the insensitivities of both Shell and the state to the plight of the people in the oil areas of the Niger Delta inclusive those of Isoko, nothing suggests that the situation would be reversed. And for the people, they can no longer wait to be exterminated. In fact, worsening poverty in the area particularly fuels protests there. How are the Isokos resisting Shell in the area?

8.7 Reactions of the Isokos to Land Crises

8.7.1 Brief Analysis of the Politics of Protest in the Oil Areas of Nigeria

Before the discussion proper, it is important to briefly revisit the political context of protest in the oil areas. As explained earlier, the current thinking of government that there are no oil communities in Nigeria takes root from the position earlier propagated by Obafemi Awolowo of blessed memory, and who was one-time Federal Finance Commissioner under the Gowon's regime, 1967-1975. As the discovery of crude oil led to the decline in the production of agricultural export crops like cocoa and rubber, Awolowo had contended that, as a national wealth and gift of nature, no community should lay any claim to crude oil. Nor should the host areas where it is explored and produced merit any special treatment in the form of fund allocation irrespective of the attendant ecological crises that it gave rise to. On the contrary, the wealth generated from crude oil should be used for the overall development of the country. Eme Awa, a scholar of federalism, had in works, also supported Awolowo's position, insisting that since crude oil is not likely to be found in all parts of Nigeria, the proceed from its production and export should be used to improve the well-being of all people and regions in the country. But there is no where in the tenets of federalism that advocated that areas where the country's major source of wealth such as crude oil is produced should be neglected as it is the case in the oil areas of the Niger Delta of Nigeria. Or are the oil communities not part of the federation? Again, why did the cocoa producing areas in then western region of Nigeria merit special attention when cocoa was a major export crop in Nigeria? These questions call into focus why protests in the oil areas will always be politicized and at best be understood by government merely as minority agitations.

That is not all as nothing suggests from the tone of the current agitations mounted by the people of the oil bearing areas that, other non-oil producing communities should not benefit from the wealth accruing from oil production and export. After all, it was from the rents accruing from oil export and production that financed post-civil war reconstruction in the 1970s, and the development of the new federal capital, Abuja. It cannot be in the spirit of federalism to get hitherto socially backward state capitals like Minna and Katsina transformed into modernity from oil proceeds while cities like Warri and Port Harcourt and their actual oil-producing rural areas suffer utter neglect. As I maintained earlier, the issue at stake is not so much with minority politics. Rather it is the character of the state and its mode of surplus extraction that are largely at the root of the crisis in the Niger Delta. For instance, it is dangerous to assume that all non-oil producing areas in the northern part of Nigeria are being developed. Minna had a face lift because of Ibrahim Babangida who was then head of the junta between 1985 and 1993. A greater part of northern Nigeria still remains rural lacking in basic social amenities like pipe-borne water. It is the cabal of oligarchies in both the northern and southern Nigeria and their clients that are the greatest beneficiaries of the oil rent. Indigenous oil prospecting companies too, have in the past four years of their operation in the Niger Delta, demonstrated wilful pollution of the environment of the host areas irrespective of whether they are in the majority or minority ethnic group.[38] Both local private capitals and the oligarchies ensure therefore, that protests in the oil areas will always fail in order to ensure that they, the chief culprits, are not brought to book.

All this accounts in part why protests in the oil areas have not been well focused. For instance, protesting groups like MOSOP took advantage of minority politics to mobilize their people into halting oil exploration and production. But, to the Nigerian state, protesting under the cloak of minority is no less than an attempt to break away from the Nigerian federation. Which explains to an extent why protests are met with military force. It is within this context that the reactions of the Isokos are examined. Now back to the real discussion, beginning with the pattern of protest in the Isoko oil-bearing area.

8.7.2 Pattern of Protests in Isokoland

As noted earlier, just as the Ogonis are not the most neglected and deprived, so, too, did they not begin protest in the Niger Delta's oil areas. For, by 1974, a year after Shell had its flowlines ruptured in Igolu at locations 13 and 18, the Isokos started a clarion call for the exit of the company from the area as a result of its degradation of their environment. Since then, no year has passed without the Isokos renewed call for Shell to quit if it would protect their environment. Yet, restive as the Isokos have been, not much impact has been achieved as Shell still operates there. Why has it been so? The pattern of protests partly explains that.

8.7.2.1 Lack of an Umbrella Organisation[39]

Unlike in Ogoniland where the MOSOP provided an initial forum for the Ogonis to redress the degradation of their environment and marginalisation, demonstrations against Shell in the Isoko area have revolved around individuals, notably Francis Okpozio. From 1974 to the early 1980s, protest was more or less initiated and led by him. With a Teacher's Grade II Certificate, and limited links to both government and Shell, Okpozio had operated with only a handful of youths. The nature and direction of protest were dependent on his own initiatives. As a local politician, he was more concerned with how he could cash in on the ecological crises caused by Shell in the area to gain political popularity than really stopping environmental pollution and its resultant effects particularly on land. As a result, agitations were limited to blocking Shell's access roads, seizing the tools of the OSCs mostly Seismography Services, and holding its workers hostage for few hours and then released. In fact,he could not secure proxy agitation with oil communities outside of the Isoko oil areas. Nor were there spontaneous protests. Lack of funding and no contact with relevant international non-governmental organisations like World Wildlife Foundation were other hindrances. In most cases, protests did not therefore, last for more than one week, and thereafter, oil operations were resumed.

So, for about 12 years that Okpozio led campaign against Shell in the Isoko area, the trend of worsening environmental underdevelopment and marginalisation of the host communities deepened. Even in 1978 and 1985 when the Western Division of Shell Nigeria located at Warri, granted him audience, he could not really articulate his demand beyond refuting the problem and asking for reparation. His asking Shell to 'bring back all that it took from Isoko area' was for influence, too vague and was ignored by Shell. In any case, these was no basis to behaviour that Shell would have acted otherwise if he made a modest request. However, it was his activism in the area that helped him to gain an elective position in then Bendel State House of Assembly of which he later became its Deputy Speaker during the second republic, 1979-1983. Again, his previous activism on environmental issues saw him into the Abacha's Constitutional Conference in 1994. For both periods, he abandoned the struggle which confirmed that most of the leaders of protest groups in the oil areas are indeed, 'conflict merchants' who are only ready to use protests to further their selfish political and economic interests, but not to reverse' the backwardness of the host oil areas.

However, as Okpozio became spent, the youths, inspired by the MOSOP, now continue with the struggle. Though a handful of the indigenes with little connections in both government and the oil companies have attempted to impose themselves on the youths as their leaders, they failed largely because of their selfish material interests. At the time of this study, the Isoko Youth Movement (IYM), and the Ijaw Youth Movement were the most popular and people-oriented group protesting against the oil TNCs, notably Shell, in the west of the Niger Delta, though its support base was still local. OMPADEC contractors also formed the Association of Contractors for OMPADEC, Isoko Chapter, an umbrella body to support both Shell and the Commission by splitting the ranks of the youths. A Chapter of the Association of Traditional Rulers in the Oil Producing Areas (ATROPA) also sprang up in Isoko land. As the Nigerian Constitution does not provide any function for traditional rulers, their fraternity with military governments enables them to get contracts from both OMPADEC and the operating companies. The ATROPA is specifically set up to weaken the solidarity of the youths. As shall soon be made clear, the emergent pattern of protests in the area since 1990 is rooted in self-centeredness of the leaders. Let us now examine legal actions taken by the people against Shell and the state.

8.7.2.2 Legal Actions

Twenty years after the Land Use Decree of 1978 had come into force, perennial protests and heightened ethnic feuds between and among the people over the ownership of lands in the rural areas of Nigeria inclusive of the oil communities, suggest that the state is finding it difficult to implement the law. For the rural people especially those living in the oil areas, they still claim full ownership of all rural lands including the minefields being operated by the oil companies. Which partly explains why the same land being mined by Shell in the host oil areas of Isokos is tilled by the inhabitants of the host oil communities, notwithstanding its infertility. It can therefore, be argued that, the Isokos, like their counterparts in other oil-bearing communities in Egbema-Oguta, Ahoada and Ogba are not actually contesting their ownership of rural lands with the state. Rather, protests in the area are concerned with the urgent pressure on Shell to either stop polluting their land or quit, though the intention has been diverted by the conflict merchants.

On its part, the state thinks otherwise, stressing that the Land Use Decree of 1978 is quite explicit on who owns the land where crude oil is found and produced. And as a result, it does not reckon with the assertion of ownership by the people over rural land. It (state) does not believe that squatters on its own land can sue the oil companies for trespass, claim damage for land, and demand compensation for crops. It is for these reasons that both the oil companies and the state do not pay compensation on land. Furthermore, since the judiciary is part of the state structure, the law court as explained, cannot be the last hope of the people in the rural oil areas in the on-going protests against the state and the oil companies over land pollution. It is within this context that the politics of legal protest over land is set. Illustrations with some specific legal cases that the people of Isokos have instituted against Shell, will shed light on the issue.

Bathed by River Ase, the Ivrogbo community has long beaches and creeks where fishing and the digging of white sharp sand constitute one of the main economic activities of the people. But this was before the advent of Shell in the area in 1959. Since the company began producing crude oil there in the past 40 years, not only were the creeks polluted which resulted in sharp decline in fishing, the large deposits of white sand were sucked away to permit laying of oil/gas pipelines. The dredged creeks became too deep for the use of small hand-dug boats both for fishing and transportation. The re-direction of the natural flow of water in the creeks away from the Ase River to facilitate oil exploration and production, was found to have disrupted the migration of fish to the area during the rainy season. It was the social and economic hardships brought about by Shell's operations as explained above, particularly the pollution of their land, creeks, and destruction of fishing and sand digging that compelled the people of Ivrogbo community to sue Shell in September 1989. They demanded for N9 million as damage, and prayed the court to halt Shell's operation in the area.[40]

In response, the presiding judge invoked the Land Use Decree of 1978 and struck out the Ivrogbos claim to land. Thereafter, the case was closed. That the judge refused to halt Shell's operations contrary to the wish of the people is not surprising. For, in a similar case that involved the Ugborikoko people and Shell, the court did not grant their prayer to stop the company's operation. The Ugborikoko people in the Warri oil area as explained earlier, had, in 1981, taken Shell to court for 25 years of flaring gas day and night, the effects of which had left their land polluted, while posing health crisis. The court, as an apparatus of the state, could not stop Shell for flaring gas since oil is the backbone of the economy of the country.[41] The case of E. Amagada versus Shell and its OSC, will throw more light on the hollowness of legal protests in the oil areas.

In 1993, E. Amagada's fish pond located within an area that later became one of Shell's block, was reported to be completely polluted by Seismograph Services Limited operating for Shell there. To him and most people in the oil areas, Seismograph Services Limited is part and parcel of Shell's network of capitalist operation. Angered by the monumental damage done to his fish pond in which dead fishes were found floating, he filed a legal suit against Shell and its OSC, claiming N200,000 as damage. As to be expected, the judge did not recognize his right of ownership of a pond in a minefield though his pond was there before Shell got the block. Nor was the judge satisfied with basis on which he computed the damage, whether on the basis of the dead fishes and at what rate, or incapacity of the pond to regain its marine life. That is not to assume that it would have awarded damage if the reasons were acceptable to the judge. At the end, the case was struck out.[42]

Though the court invoked the Land Use Decree of 1978 to nullify Amagada's case, Shell could not resume work there. For, Amagada, like most people and communities in the oil areas, had laid daring any oil worker to move into the field. As legal battles over land are lost, the people also invoke evil spiritual forces such as sending honey bees, snakes and other evil forces to chase out the oil workers. Some are even kidnapped. All is to make the host areas hostile for Shell and other oil companies to operate. So, to allow oil operations to continue, Shell would settle with the people even after the court had ruled in its favour. It was reported that Shell had offered to pay Amagada N17,000 in spite of the fact that he lost the case. As of March 1995, 24 months after it made the promise, he was yet to be paid.[43] Amagada was not alone in this kind of pranks played by Shell; for the company has reneged on its promises to the people in the Benneseide oil communities of Bayelsa State.

Because of land crises in the oil areas, previously decided land cases between and among communities are being renewed. It is a new trend of land conflicts being given rise to, as a result of acute landlessness. The fact that the court is handling such land cases in the oil areas is in part an admission of the failure of the Land Use Decree of 1978. For, if the state were the sole owner of all lands including the farmlands in the acreages over which the people are renewing claims of ownership, it would not have entertained them in the first place. The case between Egevafo Ekpeto and Ikono Wanogho was one of such recent land cases.

E. Ekepto and members of Evror family in Emevor had, after the land case was decided in June 1975 in favour of Ikono Wanogho and members of Wanogho family in Orogun under the Ughelli oil areas, renewed the legal battle in 1989. Faced with the grim reality of landlessness and its attendant crises, E. Ekpeto and others sought for legal redress to repossess the land. As of 1989, it was the only parcel of land where oil was not exploited, though seismic operations had taken place there, meaning it could still be better than other heavily polluted from land. For the second time, they lost the case, reconfirming that the land in question actually belonged to Ikono Wanogho and his family[44] But this is one of the few cases settled by the court. In other areas, land hunger is breeding violent and bloody confrontations between the parties concerned such as in the lingering clashes between the Okirikas and Ogonis. In the Isoko area too, the people of Emede and Igbide had bloody land conflict in 1991. In fact, it is a new dimension to land and ethnic feuding which is too complex to be treated in details in this study.

If the legal battle the people of the Isoko oil-bearing areas had waged and still wage against Shell and the state over land has hardly be won, it does not mean that they are no longer agitating for the protection of the environment and reversing their marginalisation. So, what other forms have protests taken in Isoko area?

8.7.2.3 General Protest

Bitter with how one F. Okpozio appeared to have dumped his own people after using them during the anti-Shell campaign to advance his own political interest; and equally angry with how some of their own indigenes are still conniving with Shell and OMPADEC to undermine their social and economic well-being, the youths have, under the umbrella of the Isoko Youths Movement, taken control of general protests against Shell, the state and their allies in the area. Compared with the MOSOP, the IYM was still faced with serious leadership crisis, finance, coordination and outreach etc. Though the IYM's numerical strength of about 500 youths is far less than that of the Ijaw Youth Movement of over 3000 youths, both organisations enjoy a reasonable land of coordination, and intelligence gathering on the activities of the state and the oil companies. In fact the IYM's main strength is its spread across the various oil communities. By 1995, the IYM had 18 chapters with each catering for a clan, and the remainder as the mobile Secretariat. Membership is open to all youths who must be indigenes, and must not have aided Shell, or other oil companies/OMPADEC to undermine agitations against environmental pollution in the past or at present. One S. Ogba who once touted himself as the leader of the Movement was disowned by the youths on grounds of alleged cases of complicity with, and gratification he received from the oil companies. At the time of this study, the Movement had a collective leadership with each clan having a nominee in the interim executive charged with organising protests, sabotage and other activities that aimed at compelling to stop its recklessness or quit the area. The most vital activity of the IYM for now, is to sensitize and mobilize the youths for action against Shell, the state and their allies there. Because the state is always fuelling intra-group crises as a plot to undermine the solidarity of the IYM as it did in the case of MOSOP, the Movement still operates in secrecy, preferring to strike when the state and Shell and their allies least expected. In no area have the covert operations of the Movement been so carefully schemed and carried out than in sabotage.[45]

Shell still suffers from one form of sabotage or the other not only in the Isoko oil areas, but across all its host communities in the Niger Delta. In cutting the oil pipelines for instance, the act is perfected and the youths ensure that the flowlines are empty in order to avoid any casualty as earlier explained. Blocking the OSCs from entering into their communities is done with the collaboration of the neighbouring oil areas. As noted, too, it is not that the youths do not understand the ecological implications of cutting oil pipes. Rather, it is an act borne out of frustration and poverty. Oil pipes therefore, are wilfully cut to ensure that both Shell and state also lose both crude oil and revenue. That the youths are able to work in alliance with some of the oil workers of Shell, is a threat to the company. Also, they that know the spots along the company's right-of-way where surveillance by the police, army and navy does not cover shows that Shell is into a tough battle. All this partly explains why nobody was, as of 1997 caught for sabotaging Shell's operations in the Isoko area an act in which

3 million barrels of crude oil had been lost in the last ten years.[46] Shell had tried without success to fish out collaborators among its staff, the main reason being that those aiding sabotage from within are local staff who are indirectly avenging Shell's wilful pollution of their own area and the company's complicity with the state in killing their brothers and sisters in Ogoniland and Ijaw oil-bearing areas. In the off-shore locations such as in Burutu and Forcados, Shell's field workers including those of its service companies are getting very apprehensive of the trend of sabotage particularly as barges suddenly develop mechanical faults mid-sea. With the help of some of the oil workers, the youths in these off-shore oil-producing communities now disguise as aids in the barges conveying tools and men to the rigs, and while mid-sea, they hold workers hostage. In fact, it is the Isoko Youth Movement that is at the root of taking Shell workers hostage in the Enweh area. Six expertriate staff of Shell were held hostage and later freed by the IYM in the area. Expensive oil tools like drilling bits valued at One million Naira were reportedly dumped into the sea all in convert protest for the degradation of their environment and to punish Shell and other oil TNCs. In both on and off-shore areas, the oil workers are now protected by patrol teams of the Nigerian Navy, army and police especially since the hanging of the Ogoni nine men in 1995. But for how long will they continue to do this? If Shell did not find it convenient to operate in Ogoni land in spite of the permanent occupation of the area by the combat troops of the Nigerian army, then the use of brute military force by the state will not quell protests in the oil areas. More details on the responses of both Shell and the state are taken up later.

Of all the strategies deployed by the youths to hold Shell to ransom in Isoko area, the halting of new exploration was the most effective way to save the land from further pollution. Stopping this first stage of finding crude oil, mean that Shell can not commence with its production. By 1992, the IYM ensured that no new exploration took place in the area. With its network across the area, all cases of further exploration were promptly reported. So, Shell and its contractor service companies were indirectly barred from further exploration. In 1992, for instance, Seismographic Services Limited defied the order of the Isoko youths not to commence any exploration in the area, and renewed seismic operations in one of Shell's old acreages in the Ilue-Ologbo/Oweh angle. Unknown to the crew members of the company, some of the youths had already taken position in the area marked for seismic shots and ready to stop any operation. Others held the rest workers hostage and seized their tools notably, tramps, seismic recorders and vehicles.

It took, Shell and their OSC a day to secure the release of their men and materials before it finally pulled out of the area and had not resumed operation there as of 1997[47]. The solidarity protest embarked upon by the youths in the neighbouring Council oil areas of Ughelli and Burutu boosted the popularity of the IYM. To Shell, its OSCs and the state, it was a new dimension to protest in the area. For instance, while the youths held the seismic party of Seismograph Services Limited in Isoko area hostage, their counterparts at Orogun had, on a tip off, stopped another group of its workers who were about to begin seismic operation in their community. As if that was not enough, others in Bomadi and Burutu areas had sent signals warning all operating oil companies to stay off their land. Because this strategy of protest is guerrilla like it is difficult for the state security personnel to predict where the youths would launch an attack next. If this mode of protest could be encouraged across the oil communities in the Niger Delta, there is every tendency that the oil companies will sooner than later, begin to rethink their mode of operations in terms of protecting the environment particularly land.

In most of the oil producing areas, protests are not really bringing the desired results because the people are pitted against each other by the combined plots of the state and the oil companies. Deep-seated distrust among the people has degenerated into ethnic skirmishes as in the case of protracted intra-ethnic conflicts between the Okirikas, Adonis and Ogonis; and the split of MOSOP into factions. All this explains why the various oil areas do not have common agenda on how to redress the ecological crises caused by the state, Shell and other oil companies in the Niger Delta.

Having recognized these problems, the IYM is probing and dealing with all those who have in the past and at present, taken undue advantage of the struggle to enrich themselves at the expense of the well-being of the Isoko people. The Movement, like its counterpart in the Ijaw oil areas, is determined to rid the community of those regarded as the fifth 'columnists'. Particularly badly wanted by the youths, are the awardees of contracts from Shell, its OSCs, and OMPADEC either directly or by proxy, got paid, and no job was done; or executed it shoddily. Agonizing over the past failures of F. Okozio, a majority of the Isokos both at home and abroad, have embraced the IYM by helping it with secret information on the beneficiaries from these shady contracts awarded by Shell and the state through OMPADEC. Though the members of the Association of Traditional Rulers in the Oil Producing Areas, and Association of Contractors for OMPADEC in the Isoko area had helped the state security forces with names of the 'radical youths' who, in turn, were harassed and arrested, the battle of cleansing the area of 'conflict merchants' and 'opportunists' still rages on. The dethronement of the *Ovie* of Uzere will shed light on this.

Uzere clan does not only have the greatest concentration of Shell's oil wells in Isoko area and therefore, the worst hit by land pollution and other ecological crises, but remains the most restive oil community in the west of the Niger Delta. Though most of the indigenes who mediated protests in the past were accused by the youths of enriching themselves in the process, none was punished. Perhaps, this explains why the *Ovie* of Uzere, Udogri I, felt that it was his turn to use the politics of protracted agitations in the area to enhance his financial fortunes. In June 1994, the Movement brought against the *Ovie*, charges of collusion with Shell/OMPADEC/Seismograph Services Limited, that have undermined the material well-being of Isoko people and their environment, particularly land. Specific charges against him were, how he spent the N30 million that he received on behalf of the community from Shell; why he was unable to tar the road expected to link Uzere with the rest part of the area after he was paid N20 million for the contract by Shell; and why he hired casual labourers for Seismograph Services Limited, an OSC to Shell whose operations were and still responsible for the degradation of their environment and marginalisation. In anger, the youths rose against him. Unable to resist the attack, and surprisingly without the backing of the state security forces, he finally fled into exile. His experience should be a lesson for others in the sense that even the state and the oil companies do abandon their agents when the corruption is too blatant to be suppressed as in the case of the Ovie of Uzere.[48] Since the Uzere episode, other traditional rulers in the other oil-bearing areas now live in fear. In fact, there is an emerging trend of the youths' hostility against these elders which is similar to the situation in Ogoniland in which four notable chiefs believed to be diverting the cause of the MOSOP were killed by the youths.

It is not in doubt that the IwYM is becoming popular among the Isokos especially as it represents the most potent force checking the excesses of Shell and their collaborators in the area. Has it been that easy? What have been the responses from Shell and the state?

8.8 Responses from Shell and the State

To begin with, what kind of response should be expected from both the state and Shell which recognize the people who live in the Isoko oil areas, like their counterparts in the oil-producing communities, as mere squatters on their own ancestral land? This question will be answered in the light of the renewed agitations in the oil areas particularly after the Ogoni episode.

Since 1990, fierce and persistent agitation across the Niger Delta have gradually made the area very unsafe for the oil TNCs to operate. The hanging of Ken Saro-Wiwa and eight others even made it worse for Shell in particular. And as the state deployment of its troops has failed to quell protest over environmental/land pollution and marginalisation, Shell and other oil companies are beginning to be forced into a re-think. Not only that. The state has understood that military option is not the answer to the crisis in the Niger delta after all. Rather, it has known the plots of the operating oil companies like Shell to hide under the cloak of the joint venture to wilfully pollute the environment of the host oil communities. There is growing mutual suspicion between the state, Shell, and other oil companies. At the times, the state deliberately leave the oil companies to face the wrath of the aggrieved people in the oil areas before intervening. How successful this strategy will be is too early to predict since the state will not like the disruption of oil production for too long a period. Analysis of the salient aspects of Shell-state responses to protests will shed light on the hollowness of their joint reactions.

8.8.1 Community Projects

To Shell and other oil companies, it is the duty of the state to provide its people in the host oil communities with social amenities like tarred roads, electricity and pipe-borne water since it collects taxes and royalties from them. Shell only provided any of these social amenities in its host oil areas in so far as they facilitates its operations. It even expended these costs as part of oil exploration and production by charging them to the joint venture arrangement. Yet, Shell would want to be solely acknowledged in such projects thereby portraying the state as not responding to the needs of its own people. Is the issue at stake really that Shell takes all the glory for a joint effort? I think not. Rather, the development problems that the people are faced with, are not tackled. It is either Shell's access roads in Isoko oil areas, like in other oil communities, are linking only its various oil and gas fields; or its fish and cassava farms. So, too, scholarships serve basic local manpower need of the company or to assist its staff. The Ogonis, for instance, were not the first people to condemn the hollowness in Shell's community development projects, the Umuechem uprising in 1989 which left over 50 people and three policemen dead, began when the people rejected the company's development projects such as providing manual garri processing machine which every household already had[49]. The pertinent question therefore is whether pipe-borne water has much relevance to the rural Isoko people who have no fertile land to farm nor creeks to fish. This questions points to one central problem, that is, unless the capacity of the environment including land to renew itself is restored, community projects will hardly reverse the underdevelopment of the oil areas. Some detailed illustration is important at this point.

In 1971, Shell extended its commercial farm project to the Isoko area. It all began in Bori, Rivers State shortly before the outbreak of the Nigerian civil war in 1967. Cited in Ellu and as an outstation of the company's main Farm Center at Agbarho, under the Ughelli South Council Area of Delta State, the farm project was meant to assist Shell workers with garri, a bye-product of cassava which means that, from the onset, the company had no plan for the people in its host areas. When the project took off, land was not scarce at that time. But to forestal future agitation, Shell got the land being used as farms as part of its confessional blocks. Located on 10 acres, the Ellu farm was one of the company's farms specifically meant to grow cassava from which "Shell garri", packed in a 2.5kg sachet, was to be produced for its staff. By 1982, the company's cassava farm disengaged from garri processing due to a combination of problems like low yields, poor returns and serious criticism from the public that it was anti-people. The same criticisms forced the company to stop the sale of the cassava tubers and stems, whose beneficiaries were those with links in Shell especially family members of its staff. It was a huge failure both in terms of being unable to supply cassava stems to the people, and halting the worsening trend of hunger in the area. By 1994, the Ellu farm became dormant.[50]

At Igbide, Shell started a fish farm about the same time the Ellu farm took off. The Igbide fish farm was overseen by the staff of the company's largest fish farm in the Niger Delta, located at Ogbe-Ijaw, Warri. Like 'Shell garri', the fishes caught from the farm were sold only to the staff of the company. Not even the fingerlings were sold to members of the public who have private fish farms in and around Igbide.[51]

From all indication, the Ellu people did not benefit from the Shell's cassava farms in terms of farm inputs, technical advice and other extension services. Nor did the people of Igbide who are noted for fishing, gain any technical assistance from Shell. Yet, the company still accused them of being ungrateful; ungrateful for projects that were meant for the welfare of staff of Shell or what?

It is true that Shell has scholarship programme for the indigenes of its host communities. However, the programme is cornered by its workers for their children and relatives in addition to a handful of the children of its allies. In 1971, the company introduced a scholarship programme at the university level not really to help the people as such, but to raise some of its engineers locally instead of recruiting expatriate staff as a means of cutting cost. By 1980, the company already had enough engineers, though it still continued with the programme perhaps only to create a pool of skilled workers from which it will recruit should the need arise. But Shell has not recruited any of the grantees of its scholarship since 1982, meaning it was meant to calm the agitated youths.

Of the 50 indigenes who got the award in Isoko area since Shell began operation there, 30 were children and relations of Shell workers, and the remainder came from the *ivies* and chiefs of the area who were its allies. About N2 million was spent in all. In other oil areas too, very few indigenes benefitted from the programme. Forty-eight got it in Ughelli, and 10 in Burutu with N1.9 million and N100,000 spent respectively.[52] Shell's secondary school scholarship is no less than one of the unwritten conditions of service for its workers. Yet, Shell sees it as community project for the oil areas. It is one of the issues Shell raised in its renewed campaign since the Ogoni episode. But it has not paid off. At a time when the people are on the brink of extinction, nobody reckons with the paltry N2 million spent by Shell to provide education for 50 sons and daughters of Isoko who were not even employed by either the government or Shell at the time this study was conducted. That explains why its proposal to build market stalls at Uzere since 1993 is still turned down by the people. Throughout Isokoland, Shell did not build even an ante-natal clinic in spite of rising maternal deaths and poor medicare there. The State-owned hospital at Oleh and Ilue-Ologbo were without drugs. The only clinic that Shell built in the wake of the growing protests across the Niger Delta, was at Irhoike under the Ethiope Council area. The out-patient clinic began operation without any doctor, but with only three staff nurses who had no drugs to dispense.[53]

On its part, the state does not want to be seen as not responding to the yearnings of the people in the rural oil areas. As a result, it began its own community projects. But like those embarked upon by Shell, they did not address land pollution, nor were they concerned with reversing rural poverty.

The Presidential Task Force which the Babangida junta had set up in 1988, could not handle the complex ecological problems and deepening rural poverty in the oil areas. Thus, in 1992, the

Babangida regime created the OMPADEC to replace it. To every regime in Nigeria since 1970 when crude oil became the mainstay of the economy, unrest in the oil areas has always been a source of worry because of its capacity to stir up large scale political and economic chaos across the country. For Babangida junta therefore, OMPADEC was created, largely as a political machinery to quell the restiveness in the Niger Delta moreso as he had plans to civilianize himself. It was the political intention of the regime that partly explained why the founding chief executive of OMPADEC, A.K. Horsefall, was an expert in state security. Horsefall had a clear mandate from the military regime of using both carrot and stick as he deemed fit in building up clientele of political supporters in the Niger Delta. Yet, the Commission was understood wrongly by many a Nigerian, to have been created to cater for the needs of the oil areas. In fact, it is in the latter context that E. Osaghae praised OMPADEC as one of the best steps being taken by the state to alleviate the poverty in the oil areas. He was mistaken no doubt, for the Commission has, indeed, deepened the misery of the people given the political motive behind its creation and operations. This was most evident in the kind of projects that it embarked upon in the oil communities, and the manner it awarded contracts. For instance, the Commission was not concerned with halting the pollution of the environment, nor did it attempt to assess the depth of destruction already done to land. It demonstrated open hostility against the youths whom it saw to be fanning the flame of agitation under the MOSOP and IYM. It aided the arrest of the members of these groups and all those seen to be opposing the transition agenda of the Babangida regime in the area. It built up a clientele of traditional rulers and contractors who became staunch supporters of the then regime of Babangida to remain in power beyond 1993. What is more, it was to these political supporters that the Commission awarded contracts. These contracts valued at about N60 billion, were therefore, taken as gratis. Hence, most of the contractors failed to execute the jobs. The board of the Commission was sacked in 1995 and a sole administrator in the person of Eric Opia, was appointed to take charge. He was later replaced by Dan Omatsola, a retired naval officer. As the Abacha regime plans to institute a panel to probe the past activities of the Commission this year, most of the contractors have fled the country. How did the Commission fare in Isoko oil communities?

In Isoko area, the Commission had no initial idea of what it wanted done for the people. This problem was rooted in the competing selfish economic interests of the conflict merchants who claimed to represent the Isokos. The same is true of other oil-bearing areas. Having understood the trend of politics in the area, OMPADEC then solicited the support of all the traditional rulers especially those in the actual oil communities. Uzere has the largest number of oil wells; as a result, the Ovie of Uzere was most favoured in the award of contracts and other forms of gratis. He was awarded contract to build market stall for the Uzere community at N20 million, got paid and no job was done. It was gathered that the people did not want such market as there were more vital development needs. He was also reported to have gotten the contract to construct the quarter of the resident engineer of the Federal Ministry of Works and Housing, but could not get paid before he fled into exile in 1994. Other *Ivies* in the area had their own share from the scores of OMPADEC's failed contracts there. The contract for the electrification of Ellu was awarded to an unknown company for N15 million; and by 1995, not even poles were erected in the town. Though OMPADEC had on records that it had awarded contracts valued at over N500 million in Isoko area with over 75 percent of the money already paid, there was hardly anything on ground to show for it. OMPADEC was on average, a big fraud throughout the oil areas with most of the culprits being the conflict merchants and state officials[54].

As can be seen from the above analysis, neither the community development projects of Shell, nor those of the state through OMPADEC, were aimed at alleviating rural poverty, and checking landlessness in the Isoko area. The situation is not different in other oil communities. Is the Clean Nigeria Associate formed in 1982 by Shell and other operating oil companies any better? How about the NDES?

8.9 Clean Nigeria Associate (CNA)/Niger Delta Environmental Survey (NDES): Checking or Worsening Land Pollution?

The Clean Nigeria Associate, CNA came into being in 1982, as a collective response by the operating oil companies and the NNPC to the spillages at the terminals. Though Shell took the initiative and invited eight other operating foreign oil companies plus the NNPC to have the CNA in place, nothing suggests that the Associate is ready to contain environmental and land pollution in the host communities.

By 1984, Halliburton of USA, which won the contract to assemble the equipment, had installed the two stations of the CNA: one at Onne overseeing other loading terminals in the East of the Niger Delta, and the other at Forcados for those in the West.

From all indications, the CNA was equipped to handle spillages in and around the terminals, particularly after Shell had lost 500,000 barrels of crude oil as a result of the loosened underneath valves in one of its tanks farms at the Forcados terminal in 1979. Why setting up the Associate to tackle spillages only at the terminals leaving both the off- and on-shore locations where most of the economic activities of the people who live there are based? Even in the off-shore oil fields, the Associate could not act when Shell had one of its off-shore rigs in Bonny sunk in 1985 killing all the crew and causing severe pollution of the environment. In the on-shore areas of which Isoko is part, the CNA also, has no plan to tackle spillages. Hence, it does not have an operational office in any of these areas; nor are there plans to take advantage of the creeks crisis-crossing these oil areas to reach them. In Isoko area for instance, there have been several cases of burst oil pipes in which million of barrels of crude oil were spilled since the CNA took off. Yet, they were all lost to the environment.

Apart from the failed CNA, Shell has initiated yet another project which, in all appearances, does not address land pollution and general degradation of the environment. Gripped by the fear of losing most of its oil installations to the recent tide of vandalization spreading across the oil communities in the Niger Delta, Shell in conjunction with the World Bank have sponsored a study to look into the environmental situation in the oil areas notably those in Bayelsa, Delta and Rivers States. About US $100 million was voted for the project. The local study group is the Niger Delta Environmental Survey (NDES), is working in conjunction with Shell's own appointed foreign consultants on the project. The NDES is basically meant to generate primary data on the oil areas. Because Shell has never relied on Nigerian scholars for final results on previous studies on the environment for fear that they will be partial, it is for Shell's foreign consultant that analyse the data and advice the company. Worst still, the NDES took off from a political ground by excluding all matters relating to land from its terms of reference. The question is: Can environmental pollution and its attendant crises be resolved as the NDES would want us to believe, without settling the root cause of the problem, which is land pollution? It is almost impossible to halt the degradation of the environment by the oil companies without addressing the land question. Another confusing aspect of the study by the NDES is the hollowness of the concept of stakeholders. Who are these stakeholders? Neither are they the rural people, nor Shell's frontmen notably some of the traditional rulers from the areas. For, they have no equity interest in the acreage. So, if there are stakeholders, they can only be Shell and the state. Even on the composition of the NDES, Shell also had its way since it is providing the funds. The choice of G.O Onosode former chairman of Cadbury, Dunlop and NAL Merchant Bank in Nigeria to head the NDES, was influenced by his link with World Bank. Connected in government, Onosode was to handpick others to assist him in carrying out the research for Shell. Although Onosode still claims that the NDES has autonomy in the discharge of its duties, it is not clear how he can operate without directives from Shell. It is what Shell wants that will finally see the light of day from its report. Amid protests, the people have developed strategies for survival. How are they coping?

8.10 Temporary Survival Strategies Adopted by the Isokos to Cope with Land Crises and their Implication for the Environment

For every step taken by the people in the oil areas including those in Isoko to cope with landlessness and its attendant crises, a whole new range of more serious environmental problems are thrown up. This is because the depth of environmental pollution of the oil area is enough to declare it a disaster zone. Yet, Shell still continues to pollute the land. In essence, the alternative survival strategies which the people have adopted are only extending the frontiers of landlessness and misery across the communities especially as their incapacities to reproduce themselves are deepening. These will become clearer as I explain of how they really cope with the situation.

Inhabitants of the Isoko oil-bearing areas have been forced by the recklessness of Shell to willingly defy their own folk knowledge of land and environmental management by poaching the swampy mangrove forest. For every hectare of the swampy forest converted into farmland, not only did it provoke more serious flooding during the rains, more of the dry lands were gradually being eaten up by the encroaching swamp. The coastal communities of Uzere, Olomoro, Igbide, Ofagbe, Ogini still remain the worst hit. Much more disappointing is the poor yields from the swamp-mangrove-turned-farmland. For one thing, the people are not ignorant of the fact that the soil of these 'new farmlands' does not necessarily support the growth of most of the staple crops notably cassava and yam. So, in most cases, not only was the growth of the crops stunted, yields were poor. Of the plantain suckers that responded to growth initially, very few were able to flower. In most cases, the suckers died shortly after planting; and of the few that survived, one out of every 100 flowered, producing small bundles in the first and only harvest. So, hunger continued to bite harder in the land.

Poaching the swampy mangrove forests also led to the indiscriminate felling of *Obecha, Abura* among the few species of trees found in the area; an act contravening the state's forestry laws. This illegal trade in timber again could not hold them for too long since the area was not endowed with enough of such forest resources. It caused deforestation and contributed to the spread of savannah in the area. Most of the rubber plantations in the area were in swamps, with the remainder in the upland areas. From folk idea on land, swampy farmlands were planted with sugar cane, water yams, and nurseries of pepper. Since these swamps had been planted with rubber trees for about 30 years, the soil was already exhausted and therefore found unfit for the growth of these crops even after the rubber plantations were converted into farmlands. The fact that most of them were gas flare sites made their soil already seriously polluted. If for anything, rubber trees hitherto unlawful to cut down were felled as fuel woods.

The trend of farming indiscriminately on patches of marginal lands particularly those found along Shell's access roads and its burrow pits was another strategy adopted to cope with land hunger. Since most of these lands were made of heaps of earth excavated from burrow pits by Shell's contractor construction companies years back, the people thought they had potentials for plant growth. Farming at this level was on small scale and more of private gardens stretching not more than 10 square meters. It was even limited to a few of the upland areas like Ozoro. Harvest was equally poor. As of 1992 it was no longer practiced because all parcels of such land were exhausted. Repetitive farming of marginal lands most of which were minefield, became widespread. Due to scarcity of farmlands, crops were planted simultaneously in the same plot after harvest. Though it used to be practiced only in virgin forests believed to have fertile soil, its application on the marginal land stemmed out of landlessness. Harvest was generally poor. For instance, by second cropping, cassava the most commonly replanted crop, bore either one or no tuber at all.

Uprooting rural people from land particularly in the manner in which the Isokos are being displaced as a result of environmental/ land pollution can be very catastrophic. And since they live off land, coping with the social and economic crises, took other forms of provisioning. Migrant farmers who were usually welcomed to the area had, since 1975, been gradually abandoned. This is because most of the retired civil servants, retrenched workers from the cities have come home to add to the population pressure on land. As noted earlier, the Isoko community is surrounded by oil-producing areas which are faced with land crises as well. There was, therefore, a trend of husbands having to migrate with their household to non-oil producing communities in Edo State, 300 kilometer north of the area at the time this study was conducted. Most of the non-oil producing areas like Orhiowon in Edo State still has relatively fertile lands that are leased to immigrant farmers for a period of three years or more. If families were migrating hoping to come back much later to build houses among other social obligations as was the case in the past, all the better. But the trend is a plan for permanent stay until death since there are no indications that landlessness will be contained as Shell and the state continue to pollute the ecology. Although it was difficult to compute the exact number of families that fled from the area in the wake of the land crises, over 50 husbands and their households were recorded in each Oweh, Ilue-Ologbo, Enweh, Igbide, Uzere, Emevor, Ogini, Ellu, Ofagbe, and Olomoro between 1993 and 1997. It should be noted that out-migration has not freed land from excessive farming since there are no fertile lands any longer. On the contrary, its implication is that more elderly people were eventually left behind in most of the areas who can hardly contribute to the infrastructural development of these communities.

As the means of income of parents were being eroded, it increasingly became difficult for them to meet social obligations of their children especially paying their school fees. A majority of the at-the-risk youths had to engage in vocations like driving side by side with schooling in order to make ends meet. Others converge at market squares and motor parks ready to sell their labour either in cash or kind to whomsoever wanted to hire them. Women and their children were not spared. Because the Isoko area has no industry, the bulk of the temporary employers of labour were few of the indigenes who were building their own houses. With a daily paid wage of N200.00 as of 1997, families were able to have one meal a day. But that was only for one week; for soon after construction, the suffering continues.

Petty trading alone has never sustained any family in the rural areas of Nigeria. If anything, it complemented farming from which the initial capital for take-off of the business was raised. Thrift societies, another source of finance for petty traders, were also badly hit since the sharp decline in the sources of income of members have made it difficult for them to continue with their membership. Petty traders most of whom were women, engaged in buying cassava tubers instead of garri from Edo State with the hope of processing them. Not only was it cheaper, it enabled them to have more bye-products like starch, tapioca, in addition to garri. Profits were calculated in foodstuff and not cash.

From all appearances, the survival strategies have not alleviated rural poverty; nor have the responses from Shell and the state helped to reduce landlessness. What are the implications of land crises for the future development of the rural (oil-producing) communities?

8.11 Concluding Remarks: Implications of the Land Crises in Isoko Community for the Future Development of Rural (Oil) Areas in Nigeria

Landlessness and its accompanying social and economic crises have thrown up a lot of theoretical and practical issues with implications for the development of rural communities in Nigeria including those in the oil areas. Land ownership, environmental policy and minority politics are some of these issues.

Like in most former colonies in Africa, the rural areas of Nigeria suffered severe social and economic backwardness under British rule. The impact of colonialism on Africa's economy is too well known to merit in another elaboration here. What is crucial to note and being emphasized here, is that, the post-colonial state in Nigeria still does not reckon with the rural areas in calculating its town and country development plans. The Nigerian state is neglecting its rural communities at a time when countries like Tanzania under then President Julius Nyerere tried to move development to the country side. Development in whatever context, cannot begin and end in the cities. Even the Western European countries like Britain and the USA in whose image Nigeria wants to look, have their rural areas developed. In these developed economies, the tendency for most people is to want to remain in the countryside where there are less urban-induced social and economic problems like pollution and congestion. Thus, the contention being made here is that, one of the reasons for the current land crises in the rural oil areas in Nigeria including those in Isoko, is that the state's philosophy of development is too urban-biased. Suffice it to note that, unless it begins to move development to the rural areas generally, the peculiar sad plight of the Isokos among other oil communities will not be heard. The oil areas need special attention, special in the sense that environmental pollution has to be checked. In essence, there is need for a well-articulated environmental policy.

Yet, till date, Nigeria has no policy on the environment. The Koko toxic waste episode in 1988 which led to the formation of the Federal Environment Protection Agency (FEPA) in that same year, still did not call for the need for one. FEPA is itself, dwarfed from birth since it is owned by the state, a state that accounts for both urban and rural pollution. For instance, most of the cement-producing companies like Benue Cement, Ashaka Cement and Sokoto Cement polluting their host rural areas have majority equity interests held by the state. So, FEPA cannot sanction them. In the oil and gas sector where environmental pollution is most severe, FEPA still cannot act since the state is in alliance with the oil companies in the wilful degradation of the ecosystem. Again, the state has no general mining policy; meaning the oil and gas industry, and solid mineral sector are being operated without any guide on how to protect the environment. A practicable environmental policy is what the country needs at the moment so that land can have the capacity to renew itself. In fact, the recent dissolution of FEPA does not mean all will be well as the state still does not want to own up to its recklessness.

Obaseki and Omotola have made it abundantly clear that the Land Use Decree of 1978 is anti-people, and should be scrapped. To that extent, I support their position. But I think that jettisoning it is not really the problem; rather, there is need for a land policy which the country does not have at the moment. For one thing, after the failure of Ujaama, it was evident that the multiple rights which the people earlier had to land, were overridden by the overbearing Tanzanian government. Tanzanians in the rural areas were displaced and made tenants in their own native land. Thus, the Issa Shivji's Committee on Land saw the need for a people-oriented land reforms, and recommended among others, that the right of rural people to their land should be restored and respected if the development of the countryside and the country at large is to be a reality[56]. The same policy, I think, can be applied to Nigeria's oil areas and other rural communities. If the people in the oil areas or any mineral-producing area for that matter, have their rights to their land recognizes by the state, by implication, they can be stakeholders in the confessional blocks. By nature of their right to land, they can check the oil companies in an attempt to stem the pollution of the environment. Precisely, this is why inhabitants of the oil communities in Texas, USA for instance, have right to the blocks being mined by the oil companies. In fact, a land policy that addresses this issue is the type being advocated.

One of the protracted crises of Nigerian politics since the period of British rule, is the continuous agitation by the people in the minority areas for reasonable access to political power and fair representation in the central government. The Tiv riot was just one of the numerous minority agitations. In the Niger Delta, the issue of minority rights is even endogenous given the history of the people. Suffice it to note that, the Ogonis are no less minority people than the Isokos and Urhobos. It can even be argued that the current level of social and economic backwardness of these areas rests partly with the viciousness that characterized minority politics in the land. For instance, the Urhobo-Itsekiri rivalry has, rather than bring the people together to use a common front in resisting the oil companies, introduced deep-seated hatred and division among them. And to be sure, Shell, Elf and Chevron have taken advantage of the split among them to continue to pollute the environment. It was the same division among the Ogoni people that split the MOSOP and its failure to agitate with one voice.

Though it is possible to deploy minority politics to settle environmental issue, I doubt if that can work in the oil areas for now. The majority ethnic groups have always cashed in on the deep agitation among the minority people to further their political interests. Genuine as the environmental cause is, the central government will always attach the clause of secession to their struggle to deflate them. When it became clear to then Abacha regime that MOSOP had mixed minority politics with environmental issues, the easier way out of the situation was taken by charging Ken Saro-Wiwa and others for murder and finally hanged them. Environmental issues should therefore, be treated on their own merit if the oil areas are to develop.

Notes and References

1. D.A. Omoweh, 'Crisis of State Capitalism and Adjustment in the Nigerian Petroleum Industry', *21st Century Afro-Review*, Vol. 2, 1, Winter 1996, pp. 113-154.
2. E.O. Osaghae, 'The Ogoni Uprising: Oil Politics, Minority Agitation and the Future of the Nigerian State'. *African Affairs*, 94, 1995, pp. 325-344; also, see C.E. Welch Jr., 'The Ogoni and Self-Determination: Increasing Violence in Nigeria', *The Journal of Modern African Studies*, 33, 4, 1995, pp. 635-649.
3. *ibid*, (Osaghae).
4. No universally accepted definition of land has emerged. This definition is, therefore, an abstraction from N.A Ollenu, *Principles of Customary land Law in Ghana*, (Birmingham: Cal. Press) 1958; J.A. Omotola, *Law and Land Rights: Whither Nigeria*? Inaugural

Lecture, University of Lagos, 1988; and *Native Land Right Act of Nigeria*, 1916; and J.C. Riddel *et al*, *The Natural Land Law of Zaire Indigenous Land Tenure in Central Bandundu*, (University of Wisconsin, Madison: Land Tenure Centre) 1987.

5. A.D. Isichei and W.W. Samford, 'The Effects of Waste Gas Flares on the Surrounding Vegetation in South Eastern Nigeria (Mimeo), 1981.
6. A. Imevbore and S. Adeyemi, 'Environmental Monitoring in Relations to Prevention and Control of Oil Pollution', (mimeo), 1981.
7. J. Oluwatimilehin, 'The Ecological Impact of the Oil Industry in the Niger Delta of Nigeria', Unpublished M.Sc. Thesis, University of Ife, 1981.
8. I.C. Amajor, 'The Ejamah-Ebubu Oil Spill of 1970: A Case History of a 14 Years Old Spills', in *The Petroleum Industry and the Nigerian Environment*, (Lagos: NNPC/Federal Ministry of Works and Housing) 1985.
9. A. Oteri, 'Groundwater Pollution Monitoring in Environmental Investigation' in *The Petroleum Industry and the Nigerian Environment*, (Lagos: NNPC/Federal Ministry of Works and Housing) 1985.
10. D.R. Denman, *The Place of Property Studies in Land Economy*, (London: London Geographical Publications) 1978.
11. N.A. Ollenu, *op.cit.*
12. J.A. Omotola, *op.cit.*
13. S. Moyo, *The Land Question in Zimbabwe*, (Harare: SAPES Trust) 1995.
14. J.S. Obaseki, cited in J.A. Omotola, *op.cit*, pp. 27-28.
15. Though the Petroleum Act of 1969 stipulates these rules, the state even on account of its materialism, cannot enforce them. See *The Petroleum Act of 1969*, Lagos, Government Press, 1970.
16. National Population Census, 1991
17. Based on the author's interviews with the indigenes in 1991, 1993 and in 1994 in both the local dialect and pidgin English before they were translated into its present form.
18 *ibid.*
19. See Colonial Laws in Nigeria, *Native Rights Act of 1956*, (Lagos: Government Press) 1957.
20. E. Ekekwe, *Class and State in Nigeria*, (Lagos: Longman) 1986.
21. See Chapter Seven of this study.
22. See *The Story of Shell in Nigeria*, Lagos, Shell Publication, 1985. Because Shell hides data on its operation, a lot of the data were generated rom the author's field trip to the actual oil area and the

company's western divisional office at Warri conducted between 1991 and 1995.

23. Based on the author's interviews with field engineers of Dowell, Flopetrol Seismograph Services and Shell at Warri in 1992.
24. *ibid*; the author has witnessed some of the operations in the Ughelli oil areas in 1990 and 1992.
25. Based on the author's field trip to Emevor, Oweh among other oil-bearing areas in April 1992. The author had also witnessed the operation of the company at Ekiugbo-Ughelli in February 1993 where similar trend of pollution was increased.
26. The author even witnessed the sharing of land in Ilue-Ologbo in April 1992. But a lot of the information on land sharing was obtained from chiefs interviewed at Ozoro, Oleh, Oweh, and even the Registrar of the High Court at Oleh in January 1995.
27. Based on the author's field trips to the Isoko oil communities.
28. For 0.8 hectares of land, 4 people are displaced. Therefore, for 900 hectares, it will be *900* x 4 =

4,500 people.

29. Based on the author's field work; for calculation, see note 28 above.
30. *ibid*; by 1988, Shell discontinued the construction of camp sites in order not to fall easy prey to the inhabitants of the oil areas who have become very aggressive.
31. Author's field work to the area in 1992, 1993.
32. See I.K.E. Ekweozor and R.J. Snowdex, 'The Studies of the Impact of Minor Oil Spill in the Upper Bonny Estuary' in *The Petroleum Industry and the Nigerian Environment*, Lagos, NNPC/Federal Ministry of Works, 1985.
33. Based on the author's field trips to the Niger Delta since 1993. See also, Amajor *op.cit.*.
34. *ibid* (Omoweh).
35. *ibid.*
36. Based on the author's field trips to the area.
37. Data were obtained from the author's field works conducted in the area between 1989 and 1997.
38. For details, see D.A. Omoweh, 'Crisis of State Capitalism'... *op.cit.*
39. Based on the author's several interviews with the officials of the Isoko Youth Movement.
40. See Ivrogbo vs. Shell, Suit no. HCO/18/90; Oleh High Court. Additional data were obtained during the author's field trip to Ivrogbo in March 1995.
41. See Chief S.O. Ubini vs. Shell, Suit no. SCK/21/81.

42. See Amagada E vs. Seismograph and SPDC, Suit no. HCO/43/93, Oleh High Court.
43. The author confirmed the information from the Registrar of Oleh High Court in March 1995 at Oleh.
44. See Egevafo Ekpeto and Co. vs. Ikono Wanogho & Co. Suit no. HCO/27/89; and HHC/56/75.
45. Based on the author's interview with one Johnson Imonieta at Oleh in March 1995. It took the identification of the author by a member of the community before Mr. Imonieta granted him audience. Even then, he was not forthcoming with more details about the Movement.
46. *ibid*; though this figure is an estimate obtained from community records, not even Shell can given an accurate figure since a large quality of the spilled crude oil would have sunk into the soil.
47. Based on the author's interviews with John Ikanaba, a staff of Seismograph Services, Prosper Ezo, a Seismic lawyer based at Ughelli, and some of the indigenes of the Ilue-Ologbo community in March 1995.
48. Based on the author's interview with A. Ekoko, Chairman, Reconciliation Committee on the Uzere Dispute in April 1995.
49. Baed on the author's checks.
50. Based on author's field trip to the area in 1994 and 1995.
51. *ibid.*
52. *ibid.*
53. *ibid.*
54. *ibid*
55. Based on the author's field trips to the area.
56. See I. Shivji, *Report of the Presidential Commission of Inquiry into Land Matters in the United Republic of Tanzania*, Uppsala, Government of Tanzania/Nordiska.

Chapter 9

Conclusion: Towards The Development of The Niger Delta

9.1 Introduction

At this stage of the study, it is important to recall that previous scholarly writings on the oil-rich Niger Delta have not really made much effort towards enhancing our understanding of the root causes, process and nature of its underdevelopment by the Nigerian state, Shell, and other foreign oil companies operating there. To be sure, the absence of a vigorous theoretical framework in these works that fully that ought to have captured the dynamics of the pillage of the Niger Delta is in part accountable for the state's formulation of wrong policies and administration of cosmetic remedies aimed at addressing the crisis of the area. The same is true of the actions of Shell, Chevron, Mobil, among other oil companies, individuals and groups from the oil-bearing areas, both local and foreign non-governmental organizations all of which seem to be concerned with reversing the underdevelopment of the Niger Delta. This obvious lacuna in writings on the Niger Delta has partly been filled by the first three chapters of this study that further analysis is unnecessary here. In this chapter therefore, I first present a brief theoretical context within which the alternative development strategy being proposed for the Niger Delta should proceed; then, I examine its domestic and external components, suggesting policy measure, and finally, I conclude.

9.2 What the Development of the Oil-Bearing Communities in the Niger Delta Should Be

The analysis of the environmental, social and economic consequences of oil exploration and production by both Shell and the Nigerian State on their host communities in the previous chapters is done in a such a way that it not only showed the nature of the underdevelopment of the oil areas, but equally suggested ways of reversing their backwardness. But, it should not be assumed that alternative development framework is unnecessary. Rather, it is crucial to have a broad philosophical context that will guide the specifities of the development of the Niger Delta. Instead of limiting the discussion to the oil communities which play host to Shell, it is thought wise to extend it areas in the Niger Delta where other foreign oil companies like Chevron, Elf, Mobil, Texaco, Agip and Pan Ocean are operating. For, Shell is not significantly different from other oil companies in terms of degrading the environment as explained earlier. I began with the theoretical framework of the development strategy.

9.2.1 Philosophical Context of the Development Strategy

More than ever before, the Ogoni uprising of 1993 has sensitized inhabitants and groups of other oil-producing areas in the Niger Delta to the need to halt the degradation of their environment, marginalization, deprivation and poverty. Both individuals and groups have called on the oil companies and the Nigerian State to redress the neglect of their communities by way of providing basic social amenities like pipe-borne water, electricity, hospitals there. Groups like MOSOP was able to internationalize the struggle against environment degradation caused by the manner in which Shell has been operating in the area in particular and across the Niger Delta generally. Others like the Isoko Youth Movement and Ijaw Youth Movement have remained ever committed to halting the degradation of their environment and deprivation by the foreign oil companies, notably, Shell, and the Nigerian state, though their activities are still largely local compared with those of the MOSOP. It is not that the demands of the Niger Delta people and these agitating group are outrageous, nor out of phase with the reality on the ground as many a Nigerian would want to believe. As noted earlier, agitations across the Niger Delta are largely community-specific, with each community more concerned with how it can get from the oil wealth. Even most of the people who claim to represent the various oil areas are basically interested in their personal gains. All is to the detriment of those who bear the real social and economic brunts of oil exploration and production. Worst still, the inhabitants and groups in the Niger Delta make demands on both the oil companies and the state quite oblivious of the root cause of the problems which they are faced with. Hence, they fight the symptoms of the problems, leaving the fundamental causes untouched. The same is even true of the responses of the state, Shell, and other oil-producing companies operating in the Niger Delta. All this makes it necessary to construct a theoretical context that will serve as a broad framework for tackling such an enormous issue like developing the oil-bearing communities in the Niger Delta.

It should be noted from the onset that, what is offered here should not be construed to mean a blueprint or an action plan for the development of the oil-producing communities in the Niger Delta. On the contrary, it is a sketch of what the theoretical context within which the specificities of the development of the oil-rich Niger Delta would be located. As a result, the theoretical framework being suggested for the development of the Niger Delta will be largely abstract, while taking note of the logic, values, principles, and historical past of the oil-bearing across and those of Nigeria at large. This is because the logic value, principles and path of development of the Niger Delta, and any other rural community in Nigeria for that matter, are not mere abstractions, but derivations from its problem. Which means that, the nature of development to be embarked upon in the Niger Delta requires a deeper understanding of the kind of problem it is faced with. That is to say that, the contemporary structure of the underdevelopment of the oil-producing areas in the Niger Delta requires specific policy and projects.

Quite many a Nigerian scholar would want to argue that, it is presumptuous to say that the Niger Delta is underdeveloped by Shell. But a field trip to the Niger Delta by such scholars will definitely bring them to terms with the state of decay in the oil areas. That alone does not guarantee a deeper understanding of the basis issues at stake. In fact, most of the scholarly writings on the Niger Delta particularly since Ogoni revolt, show glaring inadequate knowledge of the root cause(s) of the crisis in the Niger Delta. The same is true of the demands made by both individuals and groups protesting the marginalization and degradation of the environment of the Niger Delta. Perhaps, all this explains why the demands of the oil communities and the responses from the state, the Shell, and other operating oil companies in the Niger Delta, have all been efforts arrived at actualizing the kind of development that obtains in the economically advanced countries like the USA, Britain, Germany. It is not bad for the oil communities in the Niger Delta to look like these countries. Rather, there is really no mental framework to guide such development, explaining why previous efforts, by both the state and oil companies would hardly have done much to reverse the increasing incapacity of the inhabitants of the oil areas to improve their material well-being. Suffice it to note that, both the state and Shell have not defined development in the context of the logic, values, history of the Niger Delta people.

I have explained in chapters One, Two and Three of this study, the context in which development is applied here. I just want to add that, development is not synonymous with the provision of social amenities, nor economic growth, though they determine its possibility. Nor is development basically a project of the state alone as it is often constructed to mean by the state managers, some scholars of development and development agencies. Instead, development is a process and it is in that context that the development of the Niger Delta is cast. As a process, development becomes the collective enterprise of all, namely, the state, the people, and the private sector. The development of the Niger Delta therefore, should be a process by which its people can reproduce themselves socially, economically, and politically in line with their own values, choices and experience, but not to have development imposed on them by the state and foreign capital as is currently done. In fact, the kind of development being proposed for the Niger Delta should make the people who live there, its end, agent and means. That way, I think the human face will be injected into the calculus of development of the oil-bearing areas. It should also be seen as the responsibility of all, namely, the people, state and oil companies.

In casting the development of the Niger Delta in such frame, due cognizance is taken of the strategies to actualize its aim while overcoming the political factors that would militate against it. In other words, the strategy for implementing the kind of development being advocated for the Niger Delta has to be located in a specific state, social structure, culture and meaning among other crucial variables. So, too, would its development strategy be influenced by the nature of interactions, practice and politics between and among the various groups participating in the development process. What all this suggests, is that, there should be a political system that would permit a people-oriented development to take place in the Niger Delta and Nigeria at large. This takes us to the issue of democracy.

For too long, there has been growing confusion among scholars of the political economy of Africa as to whether democracy is conducive to its development or not. In the light of Nigeria's experience, while some scholars argue that the country's backwardness is in part rooted in its not being democratized, others contend that democracy would, in fact, frustrate its development, preferring military rule or authoritarian civilian rule instead. I think this confusion should not have arisen in the first place if the people are the agents, means and end of development. But a situation whereby the hegemonic class within the state structure hijacks state political power and decides for the Niger Delta people what they want, how to provide them and in what form, without recourse to them, nor their participation in the development process, would continue to reproduce the underdevelopment of the areas even in very heightening proportions. Far from being a process and the collective enterprise of all, such development will continue to have its aim tailored to meet the selfish gains of the state managers and their cronies, which is obviously to the detriment of those who really bear the brunts of oil exploration and production. This has been the underpinning political current that has shaped, and will continue to shape the state overall approach to development in Nigeria inclusive of the Niger Delta.

All I am arguing for therefore, is that, democracy has to be first enthroned in Nigeria to permit the development of the Niger Delta to take place. It is by no means an assumption that the commencement of a civilian government in Nigeria on May 29 1999, after prolonged military regime means all is well for the Niger Delta. For, rather than its substance, democracy is still largely procedural in Nigeria. By the substance of democracy, I mean that beyond elections, the ultimate political power should reside with the people and to whom their elected representatives in government must account. On their part, the people should participate in formulating government policies and executing its projects. That way, the people would become part and parcel of the development of the Niger Delta in particular, and that of Nigeria generally. Once this process of development takes root in the country, inhabitants of the oil-bearing areas in the Niger Delta would become the end of development, and naturally, their material well-being the law of development. Let us now turn to the components of the development strategy being proposed for the Niger Delta, beginning with the internal context.

9.2.2 Domestic Political and Economic Policy Framework

Without any exception, past governments have never shown genuine interest in reversing the underdevelopment of the oil-producing communities in the Niger Delta. The Obasanjo regime is not any different so far. This is all the more so since what exists in the name of a state of which the government is only its executive organ, is no less than a public social force which is not only nominal but privatized. Its privatization further renders the so-called Nigerian state incapable of rising above the conflicts of its constituent groups thereby making it a platform where these groups contest grimly for hegemonic control of state political power as it holds the key to wealth. In no other sector of the Nigerian economy has the umbriddled quest of the state managers and the political class to use state power to accumulate been so profound than in the oil and gas industry. Caring less or not at all, for the material well-being of those who live in the host communities where crude oil is produced and their environment, those in charge of the Nigerian state have demonstrated that their main concern is to corner the rent accruing from oil production and export for their personal gains.

In its present composition therefore, the Nigerian state can hardly bring about the kind of development being advocated for the Niger Delta in this study. So, the starting point is to re-organize the Nigerian state across its constituent groups. There is hardly any other alternative political environment under which the restructuring of the state can take place than democracy. As noted earlier, what Nigeria needs is the substance of democracy which implies that it is the wishes of the people that should form the main part of the basis of state development policies and projects inclusive of those on the Niger Delta. As the nation-state concept with all its artificialities has in part precipitated serious political crisis and marginalization of the Niger Delta in the course of nation-building, it is expected that under a democratic governance, the people of the oil-bearing areas should have a decisive say in how crude oil is produced and its proceeds shared. A national conference of all the nationalities of Nigeria should be an ideal forum for the agitated people of the Niger Delta to bare their minds on their pillage. Also, it should begin the actual process of democratization in the country. Part of the gains of such sovereign national conference is that, it would provide a forum for the people to decide how they would want to live together, particularly in sharing the proceeds from the production and export of the country's mineral resources like crude oil among the various geographical zones of the country. Once it is agreed among the people how crude oil, among other minerals, is to be produced and its gains shared, much in-road would have been made into redressing the deepening issues of social injustice and inequity tearing the country's far fragile fabric of political unity apart.

That is not all. As part of the restructuring of the Nigerian state under a serious democratic dispensation, those who should be in charge of the Nigerian state and its agencies, should undergo a re-orientation. A re-orientation that would prohibit them from using politics or positions in government to actualize parochial gains, but for catering for the common goods of the people whose material well-being should be the essence of development. For, if the well-being of the people including those who live in the Niger Delta becomes the end of politics in Nigeria, less stake will be placed on it and accountability will emerge as one of its major features as ultimate political power will reside with the people.

If those entrusted with the management of the Nigerian state and the political class at large, realize that the ultimate end of politics is to serve the people, while not impoverishing themselves in the course of service. Sanity would have been injected into the state's mode of surplus extraction in the petroleum industry. So, too, would the privatization of the state by the hegemonic class within the state structure through its members are able to corner the country's petroleum sector, would probably begin to disappear. It is then and only then, that one would expect serious talk to start on the democratization of the ownership of the oil blocks as a means of empowering the people living in the Niger Delta to become effective stakeholders in the development of oil. That way too, the problems of marginalization, sabotage and vandalization of oil installations would eventually become a thing of the past, as enduring conducive atmosphere for oil production would prevail.

To contend as some Nigerian scholars, policy makers, and activists, who insist that the oil-producing areas in the Niger Delta do not really require special development strategy, is to underestimate its delicate balance which oil exploration and production have upturned with no efforts to restore in sight. Worse still, to cast the persistent demand for special development for the Niger Delta particularly by the central government in the mould of minority politics shows the hollowness of the state's own conception of development. This largely explains why the recommendation of the Willink Commission of 1958 was trivialized and later jettisoned by the government. All said, the reality on the Niger Delta at the time this study was conducted, confronts one with an unsustainable environment and acute deprivation. As part of the development strategy, a special commission for the development of the oil-producing communities in the Niger Delta should be established by the state, the people and groups from the areas, and the operating oil companies.

The proposed special Commission should be involved in the formulation of policy and projects aimed at developing the oil-bearing areas. In particular, it should prepare a dossier on Shell, and other oil companies operating in the Niger Delta and their activities elsewhere in the world in order to establish the historical trend of their operation world-wide with the hope of enforcing such standard in Nigeria. It should also have a multi-disciplinary research unit whose duty, among others, would include the determination of the current level of environmental degradation in the Niger Delta, its trend, and ways to halt it; and general sensitization of the people to various forms environmental pollution such as prompt reporting of incidents of oil spillage and the outbreak of strange diseases. Rather than operate individually, the various youth movements in the Niger Delta should be integrated into an all-embracing movement groups whose duty would be to monitor the operation of the oil companies. On its membership, the head of the Special Commission, its secretary and other key officials, should, as a matter of policy, be indigenes of the oil-bearing areas in the Niger Delta. Membership should also include representatives of the state, the oil companies and states and an observer status should be granted to the relevant non-governmental organization, local and foreign that would sit in its meetings. The Commission should be empowered to either grant or refuse entry to any oil company into the Niger Delta based on its past records, notably, adherence to the standard of good oil field practice and relations with the host community. All this would help to avoid the situation in the past when neither the then Nura Imam-led Federal Government Task Force on the Oil-Producing Areas, nor its predecessor, the OMPADEC, were not really created to address the development problems of the Niger Delta. It would also avoid a repeat of the state having to create the OMPADEC and hand-pick its staff and the Obasanjo-led government bill on the Niger Delta Development Commission without consultation with the people who bear the brunts of oil production. Rather, his government merely adopted the recommendation of the Major-General Popoola's Committee on the Niger Delta and retained part of OMPADEC.

A mining policy should be formulated as a component of the broad policy objective of the development strategy for the Niger Delta. For, what have been in existence in the Nigeria's oil and gas industry, are statutory laws most of which further consolidated the accumulative base of the state and the oil companies in the sector. In fact, the majority of them were merely copied from body of international laws on the oil and gas industry without the local capacity to interpret and execute them. All this explains in part why oil companies like Shell, Chevron, Mobil can operate with relative reckless abandon and go unpunished in Nigeria, while similar offence committed in their home countries will almost cripple their business. In order to effectively check the activities of the oil companies and state agencies involved in the process of finding and mining crude oil, the mining policy should be reflective of the positions of both the people, the Nigerian state and its constituent groups on the development of the Niger Delta. It should require all operating oil companies in the country to submit to the state and the Special Commission, annual verifiable reports of how wastes generated in the process of oil exploration and production are managed, spillages handled, and programmes for gas re-injection; and present their community development projects particularly their linkages with the host communities. This way, the oil companies would be compelled to either be responsive to the needs of their host area thereby halting their enclave approach to development, or denied entry from the onset. The onerous task of compliance with the mining policy lies largely with the independence of the judiciary which is to say that, if the mining policy or any policy on the oil sector is to be enforced, the judiciary, as expected in a democracy, should have unfettered power to discharge its function without fear or favour. In fact, all should be subject to the laws of the land: the state, the people and the oil companies. By so doing, the people's lost confidence on the judiciary would be restored, and hopefully, their resort to vandalization of oil installations as they would hardly get fair judgement from the court, would start to decline.

A land policy that would restore title of ownership to the rural people should be promulgated. The Land Use Decree of 1978 is anti-people, exploitative and should be scrapped. To be abolished too, is the Land Title Vesting Decree of 1993, the reason being that it is anti-people. Under the proposed land policy individuals, families, communities and the state should own land including those where oil is found and produced, or to be explored and produced later. That way, the inhabitants of the Niger Delta, other areas where will be found, would have a decisive say in how crude oil is produced, its rent shared, and other environmental resources managed as opposed to the current situation in which those living in the oil-bearing communities are forcefully evicted from their ancestral land by state fiat. The experience of Tanzania in terms of title to land and land registration, and that of the USA with respect to right to land and the resources found therein, are quite instructive for Nigeria.

An environmental policy should be formulated. Until the dumping of toxic wastes at Koko in 1988, the Nigerian state has been indifferent to environmental pollution in the country. Even its response thereafter, particularly in ensuring the sustainability of the environment of the oil-producing areas has not been significantly different from the pre-1988 period. The fact that the state is no less guilty than the operating oil companies in polluting the environment makes it all the more so. In order to avoid the current situation whereby the state collude with foreign capital to ensure the capacity of both the ecology and economy of the host oil-producing areas to reproduce themselves while crude oil is being exploited. Within this context, the state and all operating oil companies, local and foreign, should submit annual reports of how they would manage the wastes generated from the process of oil exploration and production, and proposals on how associated gas would be handled, particularly in terms of minimizing its flaring. Apart from paying damage for polluting the environment, be it the state or the oil companies, it should be part of the environmental policy that, the operating companies would forfeit their licences as a second offender. Based on the kind of state recommended earlier, the relevant institution that would be charged with the interpretation of the environmental policy, should enjoy considerable measure of autonomy.

A petroleum policy should be enacted whose major objective would be to ensure the sustainability and economy of the oil-producing areas, the material well-being of the people who live there, and the promotion of technological development and autonomy in the country's oil and gas sector. The existing leases on the oil concession granted to the operating oil companies including their prospecting and mining licences should be revoked moreso as they were issued without any petroleum policy. For any oil company interested in Nigeria's oil sector, it should re-apply. This should be executed within the next five years. The petroleum policy being proposed here would help to check the activities of the foreign oil companies especially in compliance with gas re-injection, maintenances of the flow-lines, management of wastes, adherence to international standards of operation. Relevant government agencies like the Department of Petroleum Resources (DPR) should be more equipped to monitor and control the activities of the oil companies in the country's petroleum industry. And for the DPR to be made effective, the capital good manufacturing industries should be established in the country. Since Nigeria already has integrated steel companies, they should introduce flat steel sheet production lines whose products serve as input for the final manufacturing of oil tools and equipment. This would reduce the country's deepening technological dependence on Shell and other oil transnational.

Once these fundamental issues are adequately tackled, the task of meeting other social needs of the oil-producing areas would have been made a lot easier. Then, the social infrastructure themselves to be provided in the Niger Delta would have relevance to the people who live there. The current situation whereby the state and Shell would want to build hospital for a people who are being polluted from gas glaring day and night in the past 40 years, would no longer arise.

9.2.3 The International Context

Rather than decline, the execution of nine Ogoni environmental activists by then Abacha junta on November 1995 has fuelled more anti-Shell campaigns across the Niger Delta. Anti-Shell feelings have, since then, also extended to other foreign oil companies like Mobil, Chevron operating in the Niger Delta. One of the major implications of all this, is that the Euro-American-based parent companies of the oil transnationals operating in Nigeria's Niger Delta, home governments of these oil giants, international development agencies and non-governmental organizations, among others, that are concerned with sustainable development are getting deep and useful insights into the wreckage done to the environment of the host oil-producing communities by the foreign oil companies. Furthermore, the persistent protests in the Niger Delta have led to the drastic reduction in the total import of crude oil, which, in turn resulted in most of the foreign oil companies' inability to supply their foreign-based customers. In essence, the disruption of oil exploration and production impacts negatively on the global problems industry. At least, two policy areas at the external level are therefore, crucial for the kind of development being advocated for the Niger Delta to succeed.

First, a foreign investment code should be put in place. The policy objective of the investment code should control the inflow of foreign investments into the various sectors of the country's economy with the hope of building up the current local business class for the country's industrialization, protecting the country's economy from being monopolized by foreign capital, developing the local capacity to domesticate technology, and stemming the tide of unemployment. As part of this broad policy objective, the investment code should guide the pattern and volume of inflow of foreign capital into the country's petroleum industry. Also, details of technical cooperation with the oil companies should be spelt-out to check any form of technological dependence of Nigeria on foreign capital. This would help stop past experience of Shell in the Oloibiri field from where the company pulled out after its oil wells dried up, leaving no form of development there except capped wells and polluted environment. Shell, bent on removing traces of its activities there, also dismantled its oil pipes in the communities in the extinct Oloibiri field. The urgent need for an investment code for the country's oil and gas sector, is that, it would halt the trend of the Oloibiri experience which Shell and other oil companies have established across the Niger Delta.

Second and final, an international policy on the enforcement of standards of practice in the oil and gas sector should be formulated by the home governments of the oil transnationals in conjunction with the host state. This is all the more so, if Nigeria, among other oil-producing countries in the Third world, were to benefit from the expected gains of globalization. As part of this policy, the home governments, non-governmental organization and the relevant critical public of the industrialized countries should mount persistent pressure on the parent countries of Shell and other oil giants domiciled in their countries to ensure that the standards of practice in the industry are maintained world-wide. It is believed that such pressure would help realize sustainable human development in Nigeria and globally moreso as it occupies centre-stage in the recent discourse on world development. Such pressure could take the form of legal action or an outright sanction against any oil company that fails to comply. Since some of the home governments and NGOs in the North have equity shares in Shell and other oil giants operating in the Niger Delta, they should summon the political will to enforce the companies' compliance with the ethics of good oil field practice in the Niger Delta.

9.3 Conclusion

There is no doubt that, if Nigeria's huge reserves of hydrocarbon is properly exploited and its proceeds judiciously used that the country would not be transformed into one of the leading industrial centres in the developing world. The experience of Nigeria however, has been that its hydrocarbon resources particularly crude oil is exploited in such a way that its proceeds were diverted into the private accounts of the key hegemonic class within the state structure, but to the detriment of the host oil-producing areas in the Niger Delta in particular, and the majority of Nigeria generally. To be able to reverse this trend, and make the well-being of the people and the country the major essence of oil production, there has to be the right political system in which the ultimate power resides with the people. Invariably, Nigeria has to be a democracy and the process of its democratization put into actions. That way, people would participate meaningfully in the social, economic and industrial development of the Niger Delta and the country at large. Hopefully, the current approach of "accumulation by terror" adopted by the state, Shell and other oil companies in the Niger Delta would gradually disappear. For, though thousands of Ogonis were killed by the state military forces, Shell has not been able to resume operation in the area as of September 1990. Nor did the recent killing of the Ijaw youth in

Bayelsa help Shell to re-establish its foot hold in most of the major oil-bearing communities in the Delta region. Rather, together with the impoverishment of the communities in the old Oloibiri field, Shell's credibility across Nigeria and globally, has been on the wane. Instead of buck-passing, I think it would profit Shell to comply with the world standards of practice in the industry in Nigeria. In fact, there is need for a social contract between the state, Shell, other oil companies, and their host communities in the Niger Delta.

Without any doubt, all those who assist the state, Shell, and other oil companies to operate the underdevelopment of the Niger Delta in particular, and Nigeria generally, will oppose the development policies and strategies being proposed here. However, the increasing level of awareness on environmental protection and the resistance to marginalization by the people of the Niger Delta all signal that, now is the time for change in the Niger Delta. The collaborators of Shell and the state are left with one option: contribute to the genuine development of the Niger Delta or be consumed by the struggle.

Select Bibliography

Archival Materials

Nigerian National Archives, Ibadan

General Correspondence: Africa, Niger Coast Protectorate Drafts and Despatches 1897 - `898

F.O 462/2

C.O. 520 Southern Nigeria Despatches

C.O. 588 Proclamations and Ordinances

Lord Lugard Papers

C.O. 520/128 Confidential Enclosures

Ughelli Papers: They consist of hundreds of files on the records of the District Office, Ughelli. These papers were found very useful.

Warri Papers: Packed in files numbering hundreds. Those relevant to the study were consulted.

Bonny Papers: Also in files like those of Ughelli and Warri. Relevant ones were used as well.

Printed Books

Geary, William. *Nigeria Under British Rule*, London, 1927.

Talbot, Amaury. *The Peoples of Southern Nigeria*, Vol. 1, London 1926.

Articles

Thompson, Greame. "Some Problems of Administration and Development in Nigeria "*Journal of the Royal African Society*, xxvi, 1926-27.

Welch, James. "Witchcraft and Christianity in the Niger Delta", *Church Overseas*, IV, 1931.

Books

Adedeji, Adebayo. *Nigeria Federal Finance: Its Development, Problems and Prospects* , New York: Hutchison, 1969.

Ake, Claude. *A Political Economy of Africa*, London: Longmans, 1981.

___ ed. *Political Economy of Nigeria*, London: Longmans, 1986.

___ *Revolutionary Pressures in Africa*, London: Zed Press, 1978.

___ *Social Science as Imperialism: The Theory of Political Development*, Ibadan: Ibadan University Press, 1979.

Akinsaya, Adeoye. *Multinationals in a Changing Environment*, New York: Praeger, 1984.

Amin, Samir. *Neo-Colonialism in West Africa*, New York: Monthly Review Press, 1973.

___ *Accumulation on World Scale: A Critique of the Theory of Underdevelopment* (vols. 1 & 2), New York: Monthly Review Press, 1974.

Arrighi, Giovani and Saul, John. *Essays on the Political Economic of Africa*, New York: Monthly Review Press, 1973.

Avineri, S. ed. *Karl Marx on Colonialism and Modernisation*, New York: Doubleday, 1969.

Awa, Eme. *Federal Government in Nigeria*, Berkely: University of California Press, 1964.

Baran, Paul. *The Political Economy of Growth*, Harmondsworth: Pengiun Books, 1973.

Bierstaker, Thomas. *Distortion or Development? Contending Perspectives on the Multinational Corporation*, Mass.: MIT, 1978.

Brett., Ernest. *Colonialism and Underdevelopment in East Africa: The Politics of Economic Change, 1919-1939*. New York: Nok Publishers, 1973.

Bukharin, N. *Imperialism and the World Economy*, London: Merlin, 1970.

Clough, Raymond. *Oil Rivers Traders: Memoirs of Iboland*, London: C. Hurst, 1972.

Clower, R.W. *et al*. *Growth Without Development: An Economic Survey of Liberia*, Evaston: North-Western University Press, 1966.

Crouge, Susanne. *The World and Nigeria: The Diplomatic History of the Biafran Civil War, 1967-1970*, London: Sidnwick and Jackson, 1972.

Crowder, Michael and Ikime, Obaro eds. *West African Chiefs: Their Changing Status Under Colonial Rule and Independence*, Ile-Ife: University of Ife Press, 1970.

Dike, Kingsley. *Trade and Politics in the Niger Delta 1830-1885: An Introduction to the Economic and Political History of Nigeria*, Oxford: Claredon Press, 1956.

Dunn, John ed. *West Africans States: Failure and Promise*, Cambridge: Cambridge University Press, 1979.

Ekewe, Eme. *Class and State in Nigeria*, London: Longmans, 1986.

Ekundare, Olufemi. *An Economic History of Nigeria 1860-1960*, London: Methuen and Co. Ltd., 1973.

Emmanuel, Arghiri. *Unequal exchange: A Study in the Imperialism of Free Trade*, London: New Left Books, 1972.

Engels, Frederick. *The Origin of the Family, private Property and the State*, Moscow: Progress Press, 1974.

Evans, Peter. *Dependent Development: The Alliance of Multinationals, State and Local Capital in Brazil*, New Jersey: Princeton, 1978.

Falola, Toying. ed. *Britain and Nigeria: Exploitation or development?* London: Zed Books, 1987.

Fufunwa, Babs. *History of Education in Nigeria*, London: George Allen and Unwin, 1974.

Flint, John. *Sir George Goldie and the Making of Nigeria*, London: Oxford University Press, 1960.

Furtado, Celso. *Development and Underdevelopment*, Berkeley: University of California Press, 1964.

Frank, Andre Gunder. *Dependent Accumulation and Under Development*, New York: Monthly Review Press, 1976.

___ *Unequel Exchange: An Essay on the Social Formation of Peripheral Capitalism*, New York: Monthly Review Press, 1976.

___ *Imperialism and Unequal Development*, New York, Monthly Review Press, 1977.

Freund, Bill. *Capital and Labour in the Nigerian Tin Mines*, New Jersey: Humanities Press 1981.

Gailey, Harry. *The Road to Aba*, New York, New York University Press, 1970.

Gamble, Andres and Walton, Paul. *Capitalism in Crisis: Inflation and the State*, London: Macmillan, 1976.

Gardezi, Hassan and Rasheed, Jamil eds. *Pakistan, the Roots of Hot Dictatorship: The Political Economy of a Praetorian State*, London: Zed Press, 1983.

Gutkind, Peter *et al*. *African Social Studies: A Radical Reader*, New York: Monthly Review Press, 1977.

Havnevik Kejll, ed. *The IMF and World Bank in Africa*, Uppsala: Scandinavian Institute of African Studies, 1987.

Helleiner, G.K. *Peasant Agriculture, Government and Economic Growth in Nigeria*, Illinois: Richard Irvin, 1966.

Ikime, Obaro. *The Fall of Nigeria: The British Conquest*, London: Heinemann, 1977.

___ *Rivalry in the Niger Delta: Itsekiri-Urhobo Relations and the European Presence 1884-1936*, London: Longmans 1969.

Ihonvbere, Julius ed. *The Political Economy of Crisis and Underdevelopment in Africa: Selected Works of Claude Ake*, Lagos: Jad Publishers, 1991.

___ *Nigeria: The Politics of Adjustment and Democracy*, New Brunswick: Transaction Publishers, 1994.

___ with Shaw, Timothy. *Towards a Political Economy of Nigeria: Petroleum and Polity at the (Semi) Periphery*, Aldershdt, England: Gower, 1988.

Jalee, Pierre. *The Pillage of the Third World*, New York: Monthly Review Press.

___ *The Third World in World Economy*, New York: Monthly Review Press, 1969.

Jessup, Bob. *The Capitalist State*. Oxford: Martin Roberton, 1982.

De Kadt, E. and Williams, G. *Sociology and Development*, London: Tavistock Publications 1974.

Katz, Stephyen. *Maxism, Africa and Social Class: A Critique of Relevant Theories*, Montreal: McGill University Monograph Series, 14, 1980.

Kilby, Peter. *Industrialization in an Open Economy: Nigeria 1945-1960*, London: Cambridge Press, 1969.

Kirk-Green, A.H.M. *Lugard and the Amalgamation of Nigeria: a Documentary Record*, London: Frank Cass, 1968.

___ *Crisis and Conflict n Nigeria: A Documentary Source Book 1966-1969*, London: Oxford University Press, 1971.

Lanning, Gred, and Mueller, Marti. *Africa Undermined: A History of the Mining Companies and the Underdevelopment of Africa*, New York: Pengiun, 1979.

Lenin, Vladmir. *The Development of Capitalism in Russia*, Moscow: Progress Press 1974.

Legum Colins, Zartman William, Langdon Steven and Mytelka Lynn. *Africa in the 1980s: A Continent in Crisis*, New York: McGraw-Hill, 1979.

Leys, Collins. *Underdevelopment in Keya: The Political Economy*, London: Heinemann, 1975.

Lugard, Federick. *The Dual Mandata in British Tropical Africa*, London: Blackwood and Sons, 1922.

Luxemburg, R. *The Accumulation of Capital*, London: Routledge and Kegan, 1951.

Mabogunje, Akin. *Urbanization in Nigeria*, London: London University Press 1968.

Marx, Karl and Engels, Federick. *On Colonialism*, Moscow: Progress Press 1968.

Marx, Karl. *Grundrisee: Foundations of the Critique of Political Economy*, Harmondsmorth: Pengiun Books 1973.

___ *Capital: A Critique of Political Economy, vol. ii*, Harmondsworth: Pengiun Books 1978.

___ *Capital: Critique of Political Economy, vol. iii: The Process of Capitalist Production as a Whole*, London: Lawrence and Wishart, 1974.

___ *Capital: A Critique of Political Economy, vol i: The Process of Capitalist production*, New York: The Modern Library, 1906.

Miliband, Ralph. *The State in Capitalist Society*, London: Quarter, 1973.

Mytelka, Lynn. *Regional Development in a Global Economy: The Multinational Corporations, Technology and Andean Integration*, London: Yale University Press, 1979.

Nabudere, Daniel. *The Political Economy of Imperialism*, London: Zed Press, 1977.

Nduka, Otoni. *Western Education and Western Cultural Background*, Ibadan: OUP, 1964.

Nkrumah, Kwame. *Neo-Colonialism: The Last Stage of Imperialism*, Nw York: International Publishers, 1965.

Nnoli, Okwudiba ed. *The Path to Nigeria"s Development*, Dakar: Codesria, 1982.

Nwoke, Chibuzo. *Third World Minerals and Global Pricing: A New Theory*, London: Zed Books, 1987.

Olukoshi, Adebayo ed. *Crisis and Adjustment in the Nigerian Economy*, Lagos, Jad Publishers, 1991.

___ ed. *The Nigerian External Debt Crisis: Its Management*, Lagos: Malthouse, 1990.

Omoweh, Daniel. *The Political Economy of Shell Petroleum Development Company in Oloibiri Area, Nigeria*, Stuttgart: Bread for the World, 1989.

Onimode, Bade. *Imperialism and Underdevelopment in Nigeria*, London: Zed Books, 1982.

___ *Political Economy of Crisis in Africa*, London: Zed Books, 1989.

___ *et al. Multinational Corporation in Nigeria*, Ibadan: Shraden, 1983.

Owen, R. and Sutcliffs B, eds. *Studies in the Theory of Imperialism*, London: Longmans, 1972.

Oxaal, L., Barnett, and T. Booth, D. eds. *Beyond the Sociology of Development*, London: Routledge and Kegan, 1975.

Peter Norre and Terisa Turner, eds. *Oil and Class Struggle*, London: Zed Press, 1980.

Polyansky, F. *An Economic History: The Age of Imperialism 1870-1917*, Moscow: Progress Press 1973.

Poulantzas, K. Nicos. *Political Power and Social Classes*, London: New Left Books, 1973.

___ *Classes in Contemporary Capitalism*, London: New Left Books, 1975.

Prebisch, Raul. *Towards A Dynamic Development Policy for Latin America*, New York: United Nations 1963.

___ *The Economic Development of Latin America and Its Principal Problems*, New York: United Nations, 1950.

Sampson, Anthony. *The Seven Sisters: The Great Oil Companies and the World They Made*, London: Hodder and Stonghton, 1975.

Schalz, Ludwig. *Petroleum in Nigeria*, Oxford/Ibadan: OUP/Niser, 1969.

Seidman, An and Nera; *US Multinationals in Southern Africa*, Dar es Salam: Tanzania Publishing House, 1980.

Spero, Joan. *The Politics of International Economic Relations*, London: George Allen and Unwin, 1977.

Sweezy, Paul. *The Present as History*, New York: Monthly Review Press, 1953.

Tanzar, Michael. *The Political Economy of International Oil Companies and the Underdeveloped Countries*, London: Tempel-Smith, 1969.

Turner, Loius. *Oil Companies in the International System*, London: George Allen and Unwin, 1976.

Varga, Eugen. *Twentieth Century Capitalism*, Moscow: Progress Press, 1963.

Vernon, Raymond. *The Economic and Political Consequences of Multinational Enterprise: An Anthology*, Boston: Harvary Business School Press, 1972.

Wallersterin, Immanuel. *The Capitalist Economy*, Cambridge: Cambridge University Press, 1979.

Warren, Bill. *Imperialism: Pioneer of Capitalism*, London: NLB, 1980.

Wildstrand, Carl, ed. *Multinational Firms in Africa*, Dakar: African Institute for Economic Development and Planning, 1975.

Wilber, Charles ed. *The Political Economy of Development and Underdevelopment*, New York: Random House, 1973.

Williams, Garvian. *Nigeria: Economy and Society*, London: Rex Collings, 1976.

World Council of Churches. *Ogoni: The Struggle Continues*, Geneva, 1996.

Journals

Ake, C. `Explanatory Notes on the Political Economy of Africa. *Journal of Modern African Studies*, 14, 1, 1976.

___ "Tanzania: The Progress of a Decade". *The African Review*, 2, 1, 1972.

___ "Off to a Good Start But Dangers Await". *West Africa*, 3330, 1981.

Alavi, Hamza. "The State in Post-Colonial Societies: Pakistan and Bangladesh". *New Left Review* July - August 1972.

Amin, Samir. "Underdevelopment and Dependency in Black Africa: Origins and Contemporary Forms". *Journal of Modern African Studies*, 10, 4, 1972.

Beckman, Bjorn. "Imperialism and the National Bourgeoisis". Review of African Political Economy 22, 1981.

___ Whose State? State and Capitalist Transformation: Critique of a Kenyan Debate". Review of Africa Political Economy, 20-22, 1981.

Cardoso, Fernado Henrique. "Dependency and Development in Latin America". *New Left Review*, 74, 1972.

Emmanuel Arghiri. "Myths of Development Versus Myths of Underdevelopment". *New Left Review*, 85, 1974.

Foster-Carter, Aiden. "The Modes of Production Controversy". *New Left Review*, 107, 1978.

___ "From Rostow to Gunder Frank: Conflicting Paradigms in the Analysis of Underdevelopment". *World Development*, 4, 3, 1976.

Frayling, B. "Effect of Mining on the Economy of the Plateau Province". *Farm and Forest*, Ibadan, IV, 1943.

Gana, Aaron. "The State in Africa: Yesterday, Today and Tomorrow". *International Political Science Review*, 6. 1, 1985.

Gabriel Palma. "Dependency: A Formal Theory of Underdevelopment or a Methodology for Analysis of Concrete Situations of Underdevelopment". *World Development*, 6, 1978.

Goulbourne, H. "The Problem of the State in Back Ward Capitalist Societies: *African Development*, 1, 1981.

Gilpin, R. "The Political Economy of the Multinational Corporations". *American Political Science Review*. 70, 1976.

Hopkins, Anthony. "The Creation of a Colonial Monetary Systems: The Origins of West African Currency Board". *Journal of the Historical Society of Nigeria* III, 3, June 1967.

Kapinsky, R. "Capitalist Accumulation in the Periphery: The Kenyan Case Revisited". *ROAPE*, 17, 1980.

Kitching, G. "The Concept of Class and the Study of Africa". *African Review*, 2, 3, 1972.

___ "Modes of Production and Kenyan Dependency". *Review of African Political Economy*, 8 1977.

Lall, Sanjaya. "Is "Dependence" a Useful Concept in Analyzing Underdevelopment?. *World Development*, 11, 11/12, 1975.

Langdon Steven. "The State and Capitalism in Kenya". Review of African Political Economy", 8 January-April, 1977.

___ "The Political Economy of Dependence: Notes Towards Analysis of Multinational Corporations in Kenya". *Journal of East African Research and Development*, 4, 1974.

Leys, Collins. "The "Overdeveloped" Post-Colonial State: A Re-Evaluation". Review of African Political Economy, 5 January-April, 1976.

___ "Underdevelopment and Dependency: Critical Notes". *Journal of Contemporary Asia*, 7, 1, 1977.

McMichael, P. *et al*. "Imperialism and the Contradictions of Development". *New Left Review*, 85, 1974.

Miliband, Ralph. "The Capitalist State: Reply to Nicas Poulantzas: *New Left Review*, 60, 1970.

___ "Poulantzas and the Capitalist State". *New Left Review*, 82, 1973.

O"Brien, C.D.B. "Co-operators: Class Formation in a Senegalese Peasant Society". *Africa*, XLI, 4, 1971.

Olukoshi Adebayo. "Theoretical Approaches to the Study of Multinational Corporation in the World System". *Nigerian Journal of International Affairs*, 15, 1, 1989.

Onwuka, Ralph. "False Hopes About Technology Transfer". *ORBIS: A Journal of World Affairs*, 23, 3, 1979.

Osoba, Segun. "The Deepening Crisis of the Nigerian National Bourgeoisie". *Review of African Political Economy* 13, 1978.

Poulantzas, Nico. "The Problem of the capitalist State". *New Left Review*, 60, 1970.

Richard, Joseph. "Affluence and Underdevelopment: The Nigerian Experience". *Journal of Modern African Studies*, 16, 2, 1978.

Samoff, J. "On Class, Paradigm and African Politics". *African Today*, 29, 2, 1982.

Saul, John. "The State in Post-Colonial Societies: Tanzania". *Socialist Register*, 1974.

___ "The Unsteady State: Uganda, Obote and General Amin". Review of African Political Economy, 5, 1979.

Turner, Terisa. "The Transfer of Oil Technology and the Nigerian State". *Development and Change*, 1976.

___ "Oil Workers and the Oil Bust in Nigeria". *Africa Today*, 33, 1986.

___ "MNCS and the Instability of the Nigerian State". Review of African Political Economy, 5, 1976.

Warren, Bill. "Imperialism and Capitalist Industrialization". *New Left Review*, 8, 1973.

Williams, Garvin. "Imperialism and Development: A Critique". *World Development*, 6, (July-August) 1978.

Sklar, R.L. "The Nature of Class Domination in Africa". *The Journal of Modern African Studies*, 17, 4, 1979.

Wallerstein, Immanuel. "Class and Class-Conflict in Contemporary Africa". *Canadian Journal of African Studies*, VII, 3, 1973.

Williams, Garvin. "Imperialism and Development: A Critique". *World Development*, 6, 1978,

Zartman, William. "Europe and Africa: Decolonization or dependency?". *Foreign Affairs*, 54, 2, 1976.

United Nations Publications

United Nations. *Transnational Corporations in World Development Third Survey*, New York: United Nations Centre for Transnational Corporations, 1983.

___ *National Gas Clauses in Petroleum Arrangement*, New York: United Nation Centre for Transnational Corporations, 1987.

Conference Proceedings

NNPC, Lagos. *The Petroleum Industry and the Nigerian Environment, Proceedings of International Seminar*. Lagos: NNPC, 1981, 1983 and 1985.

Onyige, Paul, ed. *Energy, Self-Reliance and National Development*. Port Harcourt: Energy and Social Development Centre, University of Port Harcourt 1986.

Government Reports

Report of Senate Committee on Petroleum and Energy on the Tour to Oil Spillage Areas on Rivers and Bendel States. Lagos: Government Printers, February 1980.

Economic Survey of Nigeria: 1959, Lagos: Government Printers 1959.

Compensation Payment for Land Structure, Rights and Others. Benin-City: Bendel State Ministry of Lands and Housing 1969.

Central Bank Reports and Statements of Account. Lagos: CBN, 1972, 1973, 1974-1985.

Monthly Returns from Joint Venture Companies, Lagos: NNPC, 1977-1992.

Government Join Venture Participation Interests in the Oil Industry. Lagos: NNPC, 1983.

NNPC Annual Statistical Bulletin. Lagos: NNPC, 1971, 1972, 1973-1992.

NNPC Monthly Petroleum Information. Lagos: NNPC, 1977-1992.

The National Accounts of Nigeria 1959-1974. Lagos: Federal Ministry of National Planning 1975.

Ughelli High Court, Suit No; UCH/48/73, November 12, 1974. Warri High Court, Suit No. SCK/21/81, 19th July 1985. Warri High Court, suit No. IV/89/71 26th November, 1973.

Shell Publications

Community Development, Lagos, Shell, 1987.

Shell in Nigeria, Lagos, Shell, 1984.

The Story of Shell, Lagos, Shell, 1982.

Newspapers/Magazines/Others

African Concord. Lagos: Concord Newspaper Group.

Daily Times. Lagos: Daily Times Nigeria.

The Guardian (on Sunday). Lagos: Guardian Newspaper.

New Nigeria. Kaduna/Lagos: New Nigeria Newspaper.

Nigerian Herald. Ilorin: Kwara State Government.

The Nigerian Observer. Benin City: Bendel Newspaper Corporation.

The Sunday Observer. Benin City: Bendel Newspaper Corporation.

Nigerian Tide. Port Harcourt: Rivers State Government.

The Nigerian Statesmen. Owerri: Imo State Government.

Times International. Lagos: Daily Times of Nigeria.

Newswatch. Lagos: Newswatch Communication.

African Concord. Lagos: Concord Group of Newspapers.

Tell, Lagos.

Ogoni Bill of Right. November 1990.

Glossary

Selected Glossary of Terms Used in the Petroleum Industry

Associated Gas	The gas that is produced with the crude oil, water sand and inorganic matters at the well head. The separation of associated gas is done at the flow station.
Condesate	A mixture of liquid hydrocarbon at atmospheric (surface) conditions which occurs as vapour in under ground gas reservoirs.
Flaring	The disposal of gasses by burning in oil fields and refineries.
Offshore Drilling	Drilling of wells beyond the shoreline either in coastal waters or far out at sea.
Oil Field	A collection of oil and gas producing wells.
Oil Well	A borehole drilled or sunk for the purpose of searching for or producing crude oil or natural gas.
Royalty	Payment made to the host government for the right to produce oil when a discovery has been made.
Seismic Exploration	A method of prospecting for oil or gas by sending shockwaves down into the earth.
Single Buoy Mooring (SBM)	Consists of a mooring buoy anchored offshore and connected by submarine hoses to on-shore storage tank. It is a substitute for natural deep water harbour.

Sour Crude	Crude oil that has large percentage of sulphur (1% or more).
Sweet Crude	Crude oil that contains little sulphur, usually less than 1%.
Wet Crude	Crude oil that contains sediments and water.
Downstream Operations	All operations that take place in the oil industry after crude oil is produced. These activities include transportation, storage, refining and marketing.
Upstream Operations	All operations that take place in the oil industry in the course of production of crude oil such as seismic shots, exploration, drilling.
Official Selling price	The price fixed by government for crude oil or product based on their qualities at any given time.
Petroleum Profit Tax	The tax paid by the oil producing companies to government on the chargeable profit based on the sales of crude oil as laid out in the manuals of Petroleum Laws of Nigeria.
Posted Prices	Historically determined prices that are often used by oil exporting countries as basis for calculating petroleum profit taxes based on assumed market price of the particular crude oil. Thus, if an oil exporting country wishes to increase its tax revenue, government may decide to increase the posted price for crude oil rather than change its tax rate.
Refinery	A plant for refining crude oil into finished petroleum products like petrol and kerosine.
Spot Market	Any imaginary commercial area where deals on crude oil, or final petroleum products among others, are made for immediate or long term availability.

Appendices

Appendix 1: Summary of questions asked during the field work for this study

A. Questions for the inhabitants of Shell's areas of operation in the Niger Delta:

1. When did Shell Petroleum Development Company begin operating in this area?
2. Since Shell began oil exploration and production in this area, what significant contribution has the company made towards its development?
3. Would you say that the Nigeria State has been forthcoming in terms of providing social amenities for this area?
4. How many incidents of oil spillage have occurred in this area? What was the extend of damage done to the environments and socio-economic activities of the people who live here?
5. Apart from oil spillage, in what other ways have the operations of Shell affected the development of the community?
6. Were you ever paid compensation for the destruction done to your farmlands creaks as a result of gas flaring by Shell?
7. If yes, how was it paid and was the amount paid comparable to the destruction done by the company?
8. Did Shall ever embark on any development programme in your area? If yes, in what aspect of rural development and to what extent did it contribute to the material well-being of the people who bear the social and economic brunts of oil production?
9. In 1979, the Federal Government enacted the Associated Gas Re-Injection Decree (ACT). How many flare sites were in your are before and after the Act has come in force?
10. Since the early 1970s, the inhabitants of most Shell's host areas of operation in the Niger Delta have embarked on violent demonstrations almost on an annual basis ostensibly to stop the

company from damaging their environment. Is there any change in the company's mode of operation? If there is no change of attitude towards oil exploration and production in what other ways have the people registered their grievances?

11. Would you support Shell's continue operation in this area? If no, why not? Is yes, would you then say that Sehll's activities have been beneficial to this community?

B. Questions for both office and field staff of Shell Petroleum Development Company

1. The history of Shell in Nigeria has been a chequered one. Could you throw more light on its origin in Nigeria?
2. From all appearances, Shell merely supervises its `contractor'-companies in virtually all stages of its operation in the Niger Delta. Why has it been so?
3. What was Shèll's reaction to the nationalization of the BP interests in the Shell/BP joint venture in Nigeria in 1979?
4. How would you evaluate the reactions of the people who live in Shell's minefields?
5. Environmental and safety laws in the USA and Europe make it virtually mandatory for Shell to operate in those countries with utmost caution. Therefore, would you agree that the company has been reckless in its operations in Nigeria?
6. Gas flaring is an inevitable process in oil production but the extent of flaring can be minimised. Why has it not been so with Shell in Nigeria?
7. Why is it that Shell does not have a Research and Development Centre in Nigeria?
8. What was the rationale behind the company's `contract labour' in Nigeria given that labour is already cheap in the country?
9. How would you assess the company's rural/community development programme in its areas of operation?

C. Question for Government Officials

1. Would you subscribe to the view that the Nigerian state has remained a revenue collector as far as its petroleum industry is concerned?
2. As a microcosm of the Nigerian state, the NNPC has been described as a "conduit pipe for siphoning public funds into the private accounts of the managers of the state". How would you react to that?
3. How would you react to the charge that the Joint Venture Agreement has been deepening the country's technological dependence on Shell?

4. Federal government's numerous laws on oil and gas, and environmental, how would explain the relative utmost recklessness with Shell operates in Nigeria?
5. By the provision of the Joint Venture agreement, both the state, represented by NNPC, and Shell ought to share according to their participating shares, the cost of oil production, gas re-injection, compensation, among others. Is it true that the state has been a persistent defaulter?
6. Acute unemployment has been a major social problem in the Niger Delta yet Shell introduced `contract labour'. Could this not be regarded as one of the glaring collusion of the state with Shell?
7. Most of the state's laws on the petroleum industry have in the main, either consolidated its source of revenue or repressed both those who bear the cost of oil exploitation and the oil workers. Why?

Appendix 2a: List of Selected Contractors and/Service Companies to Shell in Nigeria: (Foreign)

Name of Contractor/ Service Company	Address	Nature of Business
(1) Dresser Nigeria Limited	38, Adeola Hopewell Street, Victoria Island, Lagos	Supply of Chemicals
(2) Geosciences Nigeria Limited	P271, Trans-Amadi Industrial Lay/Out, Box 5133, Port Harcourt	Wireline Services
(3) Seismograph Service Nigeria Limited	P.O. Box 3320, Lagos	Major Contractor
(4) Westminster Dredging Nigeria Limited	P.O. Box 145, Warri	Civil Dredging
(5) Whipstock Nigeria Limited	8-10, Broad Street, 13th Floor, Western House, Lagos	Rental Services-Tools
(6) Atlantic Med. Oilfield Services	P.O. Box 3760, Port Harcourt	Tubbing Inspection Services
(7) Forasol Drilling West Africa	Plot, 3110B, Lekki Penninsula Victoria Island, Lagos	Major Contractor

(8) World Wide Petroleum Services	P.O.Box 6386, Lagos	Major Contractor
(9) Africa Oilfield Supply and Services Limited	Plot 272, Trans-Amadi Industrial Layout, Port Harcourt	Supply of Materials
(10) Tidex International	P.O. Box 280, Warri	Hire of Boats
(11) Bestland and Sea Services	Plot PC43, (HFP) Nurshe, Victoria Island, Lagos	Major Contractor
(12) Bariod of Nigeria Limited	Plot 1139, Victoria Island, Lagos	Well/Mad Logging
(13) Baker Nigeria Limited	8-10 Broad Street,P. O. Box 4011, Lagos	Supply of materials/Wireline Services
(14) Westminster Offshore Nigeria Limited	P.O. Box 799, Warri	Major Contractor
(15) Camco Nigeria Limited	Plot 481,16, Festival Road,P. O.Box 5233, Lagos	Supply of materials/wireline services
(16) Cameron Iron Works Nigeria Limited	P.O.5233, Victoria Island, Lagos	Supply of materials/wireline services
(17) Dowell Schlumberger Nigeria Limited	P.M.B. 12666, Lagos	Supply of Cements/Lab. Additives
(18) Core Laboratories Nigeria Limited	P.M>B.171, Effurun, Warri	Major Contractor
(19) Flopetrol Nigeria Limited	P.O. Box 184, Lagos	Wire Line Services/Lab. Analysis
(20) Nigerian Dredging and Marine Limited	P.O. Box 361, Warri	Dredging Services
(21) Geosite Surveys Nigeria Limited	P.O. Box 1045,Port Harcourt	Survey/ Draughting Services
(22) Anadril Nigeria Limited	P.O. Box 3778, Port Harcourt	Major Contractor
(23) Ensco Tool and Supply Nigeria Limited	P.O. Box 3500, Port Harcourt	Major Contractor

(24) Ecodrill Nigeria Limited	P.O.Box 3604,Port Harcourt	Major Contractor
(25) Nisco Limited	P.O. Box 322, Warri	Mechanical Construction/ Pipeline
(26) Otis Nigeria Limited	38, Awolowo Road, Ikoyi, Lagos	WireLine Services and Supply of Spread Barge
(27) Zapata Marine Nigeria Limited	P.O. Box 502, Warri	Major Contractor
(28) Nigerian Cargo Transport Company Limited	P.O. Box 3648, Victoria Island, Lagos	Major Contractor
(29) Solus Schall Nigeria Limited	Box 230, Warri	Inspection/Diving Services
(30) Wibros	P.O.Box 1057, Victoria Island, Lagos	Major Contractor
(31) WOG allied Services Nigeria Limited	P.M.B. 1049, Warri	Major Contractor
(32) Schlumberger Nigeria Limited	P.O. Box 153, Warri	Electrical Logging Services
(33) Teco Limited	P.O. Box 1384, Warri	Minor Contractor
(34) Sea Trucks Nigeria Limited	P.M.B. 1243, Warri	Hire of Sea Trucks
(35) Remon Oil Services Limited	St. Joseph Jubile Hall, Ikeja-Lagos	Major Contractor
(36) Spencer Nigeria Limited	10 Boyle Street, P.O. Box 6269, Lagos	Consulting Engineers
(37) Trevi Foundation Nigeria Limited	P.O. Box 70621, Victoria Island, Lagos	Major Contractor
(38) Kragha and Associates	P.O. Box 4216, Marina, Lagos	Major Contractor
(39) Saipem Nigeria Limited	P.O. Box 3643, Lagos	Drilling Contractor
(40) Julius Berger	P.O. Box 3643, Lagos	Major Contractor

(41) Sedco Forex Nigeria Limited	P.M.B. 5125, Port Harcourt	Major Contractor
(42) Niger Offshore Services Limited	59, Louis Street, Lagos	Major Contractor
(43) Halliburton Nigeria Limited	38, Awolowo Road, P.O. Box 3694, Lagos	Consulting Engineers
(44) Weltrade Nigeria Limited	P.O Box 53348, Falomo, Ikoyi, Lagos	Major Contractor
(45) Geo-Group	P.O. Box 3548, Benin-City	Major Contractor
(46) HLS Nigeria Limited	Plot 75, Trans-Amadi, Port Harcourt	Major Contractor
(47) Western Atlas International	Alloh Close, Enerhen Road, Warri	Major Contractor
(48) Interfield Nigeria Limited	P.O. Box 6512, Trans-Amadi, Port Harcourt	Major Contractor
(49) Dentag Nigeria	P.O. Box 3604, Port Harcourt	Major Contractor
(50) Milpark Nigeria Limited	P.M.B. 5241, Port Harcourt	Major Contractor
(51) Acdermott Nigeria Limited	P.O. Box 94, Warri	Major Contractor
(52) Alcon Nigeria Limited	P.O. Box 4616, Port Harcourt	Major Contractor
(53) Delatre Bezons Nigeria Limited	P.O. Box 52300, Ikoyi, Lagos	Major Contractor
(54) Geophysical supplies Limited	P.O. Box 52300, Ikoyi, Lagos	Major Contractor
(55) Nigerian Dredging and Marina Limited	P.M.B. 1029, Apapa, Lagos	Major Contractor
(56) Management Information system Company Limited	P.O. Box 53386, Ikoyi, Lagos	Major Contractor
(57) Haven Nigeria Computer Company Limited	P.O. Box 53027, Lagos	Major Contractor
(58) Netherland Harbour-works Nigeria Limited	Chief Okupe Estate, Maryland, Lagos	Major Contractor

(59) Data Sciences Nigeria Limited	P.O. Box 6352, Lagos	Major Contractor
(60) Soimi Nigeria Limited	P.O. Box 52781, Lagos	Major Contractor
(61) Noble Bawden Drill Nigeria Limited	P.M.B. 5218, Port Harcourt	Major Contractor
(62) Villisco Nigeria Limited	P.M.B. 2140, Ikoyi, Lagos	Major Contractor
(63) United Geophysical Nigeria Limited	73A, Timba Road, Palmgroove Estate, Lagos	Major Contractor
(64) Global Engineering Company Limited	24, Adeola Odekun, Victoria Island, Lagos	Major Contractor
(65) Prakla Seismic	5, Creek Road, Apapa, Lagos	Major Contractor
(66) Petrolog Limited	P.O. Box 1859, Lagos	Major Contractor
(67) Tristar Nigeria Limited	P.O. Box 855, Port Harcourt	Major Contractor
(68) Equator Digital System Limited	P.O. Box 6544, Port Harcourt	Major Contractor
(69) Dubi Nigeria Limited	Port Harcourt-River State	Major Contractor
(70) Telnet Nigeria Limited	1, Engineering Close, National Engineering Centre, Victoria Island, Lagos	Major Contractor
(71) Emmal Nigeria Limited	P.O. Box 852, Warri	Major Contractor
(72) Overseas Technical Services Nigeria Limited	P.O. Box 4998, Lagos	Major Contractor
(73) Petro-Dynamics Nigeria Limited	P.O. Box 6544, Port Harcourt	Major Contractor

Note: Shell changes its list of contractors/oil service companies every year. The major contractors are however, almost retained.

Sources: - Joint Venture Dept., of NNPC, Lagos
- Author's Field work 1989 - 1998
- Department of Petroleum Resources, Lagos
- Shell, Lagos

Appendix 2b: List of Selected Contractors and Service Companies to Shell in Nigeria (Local)

Name of Contractor/ Service Company	Address	Nature of Business
1. Oriental Hotels and West Limited	P.O. Box 453, Warri	Hotel Accommodation and catering services
2. Erco and Sons Nigeria Limited	P.O. Box 73, Effurun, Warri	Driver Trainer
3. Mofor Motors Limited	P.O. Box 398, Warri	General Suppliers
4. Omabikimi Opubor and Sons	96, Okere Road, Warri	Cleaning/Printing services
5. John Kinsley	7, Arighe Street, P.O. Box 122, Warri	Grass/Bush cleaning
6. Etugbo Orukkele	Emede Town, Via Oleh	Grass cutting/ Timber Handling
7. Vani International Security Limited	25, Adeniyi Jones, Ikeja,Lagos	Security services
8. F. Omo Oketete Construction Company Limited	P.O. Box 907, Warri	Supply of Labour
9. Idama Hotels	P.O. Box 2055, Warri	Supply of Labour
10. Goldmine Finance Limited	P.M.B. 1211, Warri	Hotel Accommodation
11. Joe Edah and Sons	Ogulagba, Via Warri	Road Repair
12. P.E. Asagba Enterprises	P.O. Box 75, Warri	General Services
13. Kagbo Industrial Enterprises Limited	P.O. Box 6, Warri	Printing Services
14. Ice and Tee Industrial Limited	Box 1526, Warri	Supply of Furniture
15. Pofors Bookshop	15, Warri/Sapele Road, Warri	Supply of Stationeries
16. Pat Umukoro	P.O. Box 55, Ughelli	General Services

Note: Shell has over 500 contractors servicing the company as of 1997. It is not the number that is important here, rather, the nature of their business which is menial compared to their foreign counterparts.

Sources: - Shell, Lagos
- Author's Field Work - 1989-1999
- Department Petroleum Resources, Lagos

Index